ISNM 59:
International Series of Numerical Mathematics
Internationale Schriftenreihe zur Numerischen Mathematik
Série internationale d'Analyse numérique
Vol. 59

Birkhäuser Verlag
Basel · Boston · Stuttgart

Numerical Methods of Approximation Theory, Vol. 6

Workshop on Numerical Methods of Approximation Theory
Oberwolfach, January 18–24, 1981

Numerische Methoden der Approximationstheorie, Band 6

Tagung über Numerische Methoden der Approximationstheorie
Oberwolfach, 18.–24. Januar 1981

Edited by
Herausgegeben von

L. Collatz, Hamburg
G. Meinardus, Mannheim
H. Werner, Bonn

1982

Birkhäuser Verlag
Basel · Boston · Stuttgart

Editors/Herausgeber

Prof. Dr. L. Collatz
Universität Hamburg
Institut für Angewandte Mathematik
Bundesstrasse 55
D-2 Hamburg 13

Prof. Dr. G. Meinardus
Fakultät für Mathematik und Informatik
Universität Mannheim
Seminargebäude A5
D-6800 Mannheim

Prof. Dr. H. Werner
Institut für Numerische und Instrumentelle Mathematik
Westfälische Wilhelms-Universität
D-4400 Münster

Library of Congress Cataloging in Publication Data

Main entry under title:
Numerical methods of approximation theory.
(International series of numerical mathematics ; v. 59)
English and German.
Includes index.
1. Approximation theory--Addresses, essays, lectures. I. Collatz, Lothar, 1910- . II. Meinardus, Günther, 1926- . III. Werner, Helmut, 1931- IV. Series.
QA297.5.N85 511'.4 81-21589
ISBN 978-3-0348-7188-4 AACR2

CIP-Kurztitelaufnahme der Deutschen Bibliothek

Numerical methods of approximation theory
= Numerische Methoden der Approximationstheorie / Workshop on Numer. Methods of Approximation Theory. - Basel; Boston; Stuttgart: Birkhäuser
Bd. 1-4 u. d. T.: Numerische Methoden der Approximationstheorie
Vol. 6. Oberwolfach, January 18-24, 1981. - 1981
(International series of numerical mathematics; Vol. 59)
ISBN 978-3-0348-7188-4
NE: Tagung über Numerische Methoden der Approximationstheorie; PT; GT

Softcover reprint of the hardcover 1st edition 1982

ISBN 978-3-0348-7188-4 ISBN 978-3-0348-7186-0 (eBook)
DOI 10.1007/978-3-0348-7186-0

VORWORT

Der Band enthält Manuskripte zu Vorträgen, die auf einer von den Herausgebern geleiteten Tagung über "Numerische Methoden der Approximationstheorie" am Mathematischen Forschungsinstitut Oberwolfach in der Zeit vom 18.-24. Januar 1981 gehalten wurden. Das Spektrum der Vorträge reichte von der klassischen Approximationstheorie über mehrdimensionale Approximationsverfahren bis hin zu praxisbezogenen Fragestellungen. Zu den zuerst genannten Gebieten gehörten z.B. die Verfeinerung von Fehlerabschätzungen bei der Polynominterpolation, Fragen zur Eindeutigkeit, Charakterisierung optimaler Interpolationsprozesse und Algorithmen zur rationalen Interpolation. Bei den weiteren genannten Gebieten spiegelten zahlreiche Vorträge das steigende Interesse an der mehrdimensionalen Interpolation, insbesondere mit verschiedenen Arten von Splines wider. Hier standen u.a. Probleme der Parameterschätzung in der Medizin und Flugtechnik, Fragen der Approximationstheorie bei der Konstruktion von Plottern und stabile Algorithmen beim Arbeiten mit mehrdimensionalen B-Splines im Mittelpunkt des Interesses.
Die Tagung lieferte einen repräsentativen Ueberblick über die aktuellen Trends in der Approximationstheorie.
Zum guten Erfolg der Tagung trug wie immer die hervorragende Betreuung durch die Mitarbeiter und Angestellten des Instituts sowie das verständnisvolle Entgegenkommen des Institutsdirektors, Herrn Professor Dr. Barner, bei.
Unserer besonderer Dank gilt dem Birkhäuser Verlag für die wie stets sehr gute Ausstattung.

Lothar Collatz	Günther Meinardus	Helmut Werner
Hamburg	Mannheim	Bonn

INDEX

STRENGE EINDEUTIGKEITSKONSTANTEN UND FEHLERABSCHÄTZUNGEN BEI LINEARER TSCHEBYSCHEFF-APPROXIMATION

Hans-Peter Blatt

Strong unicity constants play an important role for estimating the accuracy of given approximates for the best approximation. In this paper the asymptotic behaviour of these strong unicity constants is considered in the polynomial case. Moreover the concept of H-sets is generalized in such a way to get error estimates in the function space even if the Haar condition fails.

Bei der Fehlerabschätzung für Minimallösungen spielen strenge Eindeutigkeitskonstanten eine wichtige Rolle. In dieser Arbeit wird das asymptotische Verhalten dieser strengen Eindeutigkeitskonstanten im Polynomfall untersucht. Außerdem wird das Konzept der H-Mengen so erweitert, daß Fehlerabschätzungen im Funktionenraum gewonnen werden, auch wenn die Haarsche Bedingung nicht erfüllt ist.

1. Einführung.

Wir betrachten das klassische lineare Tschebyscheff-Approximationsproblem:

Sei B ein kompakter Raum, C(B) die Menge der auf B stetigen reellwertigen Funktionen, versehen mit der Tschebyscheff-Norm $\|\cdot\|$. Zu einem n-dimensionalen Unterraum $V \subset C(B)$ und $f \in C(B)$ sucht man eine Minimallösung $v^* \in V$, d.h.

$$\|f-v^*\| = \min_{v\in V} \|f-v\| .$$

Existiert eine Zahl $\gamma > 0$ mit

$$\|f-v\| \geq \|f-v^*\| + \gamma \|v-v^*\| \tag{1}$$

für alle $v \in V$, so spricht man von <u>strenger Eindeutigkeit</u> der Minimallösung v^*. Ist V ein Haarscher Unterraum, so ist

$$\gamma = \gamma(f) > 0 \quad \text{für alle} \quad f \in C(B) .$$

Generell (auch für $\gamma = 0$) kann man $\gamma(f)$ für $f \notin V$ folgendermaßen charakterisieren (Bartelt, McLaughlin [1]):

$$\gamma(f) = \min_{\substack{v\in V\\ \|v\|=1}} \max_{x\in E_n(f)} \sigma(x)v(x) . \tag{2}$$

Hierbei ist $\sigma(x) = \operatorname{sgn}(f-v^*)(x)$ und $E_n(f) = \left\{x \in B \mid |(f-v^*)(x)| = \|f-v^*\|\right\}$ die Extremalpunktmenge der Fehlerfunktion $f-v^*$.

Eine entsprechende Ungleichung wie (1) für Näherungslösungen hat Schaback [11] benutzt, um Abschätzungen für die exakte Minimallösung zu gewinnen:

Sei $f \notin V$ und $T = \{x_1, x_2,\ldots,x_m\} \subset B$ eine H-Menge zur Fehlerfunktion $f - \hat{v}$ (Collatz [5]). Dann gibt es Zahlen $\lambda_i \geq 0$ mit $\sum_{i=1}^{m} \lambda_i = 1$ und $\sum_{i=1}^{m} \lambda_i \operatorname{sgn}(f-\hat{v})(x_i)\cdot v(x_i) = 0$ für alle $v \in V$.
Außerdem ist

$$\|f-v^*\| \geq \min_{x\in T} |(f-\hat{v})(x)| + \gamma\|\hat{v}-v^*\|_T \tag{3}$$

mit

$$\gamma = \frac{\lambda}{1-\lambda} , \quad \lambda = \min_{1\leq i\leq m} \lambda_i , \tag{4}$$

$$\|g\|_T = \max_{x \in T} |g(x)| \quad (g \in C(B)). \qquad (5)$$

Falls also die H-Menge T minimal ist, so gilt $\gamma > 0$ und man erhält aus (2) eine Abschätzung für die Differenz $\hat{v} - v^*$ bezüglich der diskreten Halbnorm $\|\cdot\|_T$ über der H-Menge T. Falls $\|\cdot\|_T$ eine Norm ist und man die Äquivalenzkonstanten zwischen $\|\cdot\|_T$ und $\|\cdot\|$ kennt, so erhält man aus (2) eine Abschätzung für $\|\hat{v} - v^*\|$ selbst (Schaback [11]). Erfüllt V nicht die Haarsche Bedingung, so hat man oft H-Mengen, die nicht zu diskreten Normen führen bzw. mehr und mehr ausarten. Im letzten Teil wird versucht, auch für diesen Fall praktikable Abschätzungsmethoden anzugeben.

2. Abschätzung der strengen Eindeutigkeitskonstanten im Polynomfall.

Vom numerischen Standpunkt sind wegen (3) und (1) möglichst große γ günstig. Poreda [9] stellte die Frage, wie sich die strengen Eindeutigkeitskonstanten $\gamma_n(f)$ für festes $f \in C[a,b]$ für wachsendes n verhalten, wenn man bezüglich $V = \Pi_n$, dem Raum der reellen Polynome vom Grad $\leq n$, approximiert. In den letzten Jahren haben sich eine Reihe von Arbeiten damit befaßt, aus der Struktur der Extremalpunktmenge $E_n(f)$ Rückschlüsse auf das asymptotische Verhalten von $\gamma_n(f)$ zu ziehen ([2],[7],[12]).

Insbesondere existiert eine Vermutung von Henry, Roulier [7], ob

$$\underline{\lim}_{n\to\infty} \gamma_n(f) = 0 \text{ oder } f \text{ ein Polynom ist.} \qquad (6)$$

Wäre diese Vermutung falsch, dann gäbe es eine Funktion, die sich für die numerische Berechnung der Minimallösung unabhängig von n als außerordnetlich stabil verhalten würde. Die Untersuchung dieses Problems ist wesentlich verknüpft mit der Gestalt der Extremalpunktmenge $E_n(f)$. Besteht $E_n(f)$ lediglich aus einer (n+2)-punktigen Alternanten,

$$E_n(f): x_o < x_1 < x_2 < \ldots < x_{n+1} \quad ,$$

und betrachtet man die strenge Eindeutigkeit nur für die diskrete Approximation auf $E_n(f)$, so folgt aus (4):

$$\lambda \leq \frac{1}{n+2} \text{ und } \gamma \leq \frac{1}{n+1} \; .$$

Da die Zahl $\gamma_n(f)$ allein durch das Verhalten der Fehlerfunktionen auf $E_n(f)$ bestimmt wird (Henry, Roulier [7]), so folgt

$$\gamma_n(f) \leq \frac{1}{n+1} \; . \qquad (7)$$

Es gilt somit folgender

<u>Satz 1 (Schmidt [12]):</u> Existiert zu f eine Teilfolge $\{n_k\}$, so daß $E_n(f)$ genau aus n+2 Punkten besteht für $n = n_k (k=1,2,\ldots)$, so gilt:

$$\underline{\lim}_{n\to\infty} \gamma_n(f) = 0 \; .$$

Falls $E_n(f)$ aus mehr als n+2 Punkten besteht, zerlegt man $E_n(f)$ in endlich viele Vorzeichenkomponenten

$$E_n^o < E_n^1 < \ldots < E_n^{m+1}$$

mit den Eigenschaften:

$$\sigma(x) = \mathrm{sgn}(f-T_n(f))(x) \text{ ist konstant in } E_n^i\,, \qquad (8)$$

$$m \text{ ist minimal.} \qquad (9)$$

$T_n(f)$ ist hierbei die Minimallösung zu f bezüglich Π_n.

Bemerkung: Ist $T_n(f) \neq T_{n+1}(f)$, so ist m = n und alle möglichen Alternanten von f - $T_n(f)$ erhält man, indem man jeweils einen Punkt aus E_n^i auswählt.

Wir betrachten nun folgende Aufgaben (A_i):

Zu fixiertem $y_i \in E_n^i$ bestimme man $p_i \in \Pi_n$ so, daß

$$p_i(y_i) = -\sigma(y_i)\,, \qquad (10)$$

$$\sigma(x)p_i(x) \le h_i \text{ für alle } x \in E_n(f)\,, \qquad (11)$$

$$h_i \text{ minimal ist.} \qquad (12)$$

Die Lösungen dieser Aufgaben werden folgendermaßen charakterisiert.

Lemma: p_i ist Minimallösung zur Aufgabe (A_i) genau dann, wenn Punkte $x_k \in E_n(f)$ ($k = 0,\dots,n+1$, $k \neq i$) existieren mit

$$x_o < x_1 < \dots < x_{i-1} < y_i < x_{i+1} < \dots < x_{n+1}\,, \qquad (13)$$

$$f(x_k) - T_n(f)(x_k) = \sigma(x_k) \quad \text{für } k \neq i\,, \qquad (14)$$

$$\sigma(x_k)\cdot p_i(x_k) = h_i \quad \text{für } k \neq i\,, \qquad (15)$$

$$\sigma(x)\cdot p_i(x) \le h_i \quad \text{für alle } x \in E_n(f)\,. \qquad (16)$$

Bemerkung: Auf Grund dieses Lemmas sind die p_i eindeutig bestimmt. Außerdem kann man die p_i mit dem Remez-Algorithmus (mit einer Interpolationsbedingung bei y_i) berechnen. Die Charakterisierung von $\gamma_n(f)$ durch (2) führt in Verbindung mit dem Lemma zu

Satz 2: $\gamma_n(f) = \min_{1\le i\le m+1} (h_i/\|p_i\|)$.

Diese neue Charakterisierung läßt eine Verschärfung von Satz 1 zu, nämlich

Satz 3: Sei $f \in C[a,b]$ und $\{n_k\}$ so gewählt, daß $T_{n_k}(f) \neq T_{n_k+1}(f)$. Weiterhin sei $m(n_k)$ die Anzahl der $E^i_{n_k}$, die nicht einpunktig sind. Ist $m(n_k) = o(\log n_k)$ für $k \to \infty$, so gilt

$$\underline{\lim}_{n\to\infty} \gamma_n(f) = 0 .$$

Folgerung: Satz 3 gilt für $m(n_k) \le m_o$ für alle k. Insbesondere enthält Satz 3 ein Ergebnis von Bartelt, Schmidt [2], wo $m_o = 2$ gesetzt wurde.

3. A posteriori-Fehlerabschätzungen für die Minimallösung.

Wir untersuchen Fehlerabschätzungen der Typs (2) und betrachten Erweiterungen der H-Mengen von Collatz [5], wie sie in neueren Algorithmen vom Remez-Typ vorkommen.

Definition: R heißt eine H-Menge bezüglich V, wenn $R \subset S_{V*}$ und der Nullpunkt von V* in der konvexen Hülle von R liegt.

Hier ist V* der zu V gehörende Dualraum und S_{V*} die Einheitskugel des Dualraums.

Satz 4: Es sei $R = \{L_1, L_2, \ldots, L_m\}$ eine H-Menge und $\sum_{i=1}^{m} \lambda_i L_i \in V^\perp$, $\lambda_i \ge 0$, $\sum_{i=1}^{m} \lambda_i = 1$. Dann gilt für die Minimallösung v* zu f bezüglich V und für jedes $v \in V$

$$\|f-v^*\| \ge \min_{1\le i\le m} L_i(f-v) + \gamma\|v-v^*\|_R \tag{17}$$

mit $\gamma = \lambda/_{1-\lambda}$ und $\lambda = \min_{1\le i\le m} \lambda_i$.

Dabei verstehen wir unter $\|\cdot\|_R$ die diskrete Halbnorm

$$\|v-v^*\|_R = \max_{1\le i\le m} |L_i(v-v^*|$$

und unter $V^\perp$ das orthogonale Komplement zu V in V^*.

<u>Beweis:</u> Die Behauptung kann man analog zur Ungleichung (3) beweisen (vgl. Blatt, Klotz [3], Schaback [11]):

Es ist

$$L_i(f-v^*) = L_i(f-v) + L_i(v-v^*) ,$$

also

$$\|f-v^*\| \ge \min_{1\le i\le m} L_i(f-v) + \max_{1\le i\le m} L_i(v-v^*) .$$

Gibt es ein j mit $L_j(v-v^*) = \|v-v^*\|_R$, so ist die Behauptung bereits erfüllt. Andernfalls existiert ein j mit

$$L_j(v-v^*) = - \|v-v^*\|_R .$$

Wegen $\sum_{i=1}^{m} \lambda_i L_i \in V^\perp$ folgt

$$0 = \sum_{i=1}^{m} \lambda_i L_i(v-v^*) = -\lambda_j \|v-v^*\|_R + \sum_{i\ne j} \lambda_i L_i(v-v^*)$$

oder

$$\|v-v^*\|_R = \frac{1}{\lambda_j} \sum_{i\ne j} \lambda_i L_i(v-v^*) \le \frac{1-\lambda_j}{\lambda_j} \max_{1\le i\le m} L_i(v-v^*)$$

$$\le \frac{1-\lambda}{\lambda} \max \lambda_i(v-v^*) .$$

Damit ist die Behauptung bewiesen.

Vom praktischen Standpunkt sind solche H-Mengen wichtig, die zu einem positiven λ und zu einer Norm $\|\cdot\|_R$ bezüglich V führen. Betrachtet man in V eine Zerlegung in eine direkte Summe

$$V = [R]^{\perp} \oplus W$$

so ergibt sich für die entsprechenden Zerlegungen

$$v = v_1 + w_1 \text{ , } v^* = v_2 + w_2$$

mit $v_i \in [R]^{\perp}$ und $w_i \in W (i=1,2)$ mit den Bezeichnungen des vorigen Satzes aus (17):

$$\|f-v^*\| \geq \min_{1\leq i\leq m} L_i(f-v) + \gamma \|w_1-w_2\|_R \text{ .} \tag{18}$$

Hier ist dann $\|\cdot\|_R$ eine <u>Norm</u> im Unterraum W. Man kann also zu einer Näherung V abschätzen, um wieviel sich die Anteile w_1, $w_2 \in W$ für die Näherung v und die Minimallösung v* unterscheiden.

Um zu einer Abschätzung für den gesamten Raum V zu gelangen, führte Schaback H_s-Mengen ein. Wir wollen hier eine Modifizierung dieser H_s-Mengen vorschlagen, die folgende Vorteile besitzt: Einmal treten solche Mengen in natürlicher Weise bei Algorithmen vom Remez-Typ auf (Carasso, Laurent [4], Ruffer-Beedgen [10]), und es wird weitgehend für die Fehlerabschätzung weiterer Rechenaufwand vermieden. Zum anderen führt diese Modifizierung zu schärferen Fehleraussagen.

<u>Definition:</u> $R \subset S_{V^*}$ heißt H_s-Menge, wenn dim R = n ist und zu jedem $L_i \in R$ eine Zahl $0 < \lambda_i < 1$ und ein $L_i^{\sim} \in \text{conv}(R)$ existiert mit

$$\lambda_i L_i + (1-\lambda_i) L_i^{\sim} \in V^{\perp} \text{ .} \tag{19}$$

Satz 5: $R = \{L_1, L_2, \ldots, L_m\}$ sei eine H_s-Menge. Dann gilt die Ungleichung

$$\|f-v^*\| \geq \min_{1\leq i\leq m} L_i(f-v) + \gamma\|v-v^*\|_R$$

mit $\gamma = \min(1, \frac{\lambda}{1-\lambda})$, $\lambda = \min_{1\leq i\leq m} \lambda_i$ und die Zahlen λ_i sind die Zahlen aus (19).

Beweis: Es ist

$$L_i(f-v^*) = L_i(f-v) + L_i(v-v^*)$$

und

$$\tilde{L}_i(f-v^*) = \tilde{L}_i(f-v) + \tilde{L}_i(v-v^*).$$

Somit gilt:

$$\|f-v^*\| \geq \min(L_i(f-v), \tilde{L}_i(f-v)) + \max(L_i(v-v^*), \tilde{L}_i(v-v^*)) \qquad (20)$$

Wegen (19) ist

$$\tilde{L}_i(v-v^*) = -\frac{\lambda_i}{1-\lambda_i} L_i(v-v^*). \qquad (21)$$

Also ist sgn $\tilde{L}_i(v-v^*) = -$ sgn $L_i(v-v^*)$. Falls nun $L_i(v-v^*) \leq 0$ ist, so folgt:

$$\tilde{L}_i(v-v^*) = \frac{\lambda_i}{1-\lambda_i} \left|L_i(v-v^*)\right|$$

und

$$\max(L_i(v-v^*), \tilde{L}_i(v-v^*)) \geq \left|L_i(v-v^*)\right| \min(1, \frac{\lambda_i}{1-\lambda_i}) \qquad (22)$$

Da außerdem $\tilde{L}_i \in \text{conv}(R)$ ist, gilt

$$L_{\tilde{i}}(f-v) \geq \min_{1\leq j\leq m} L_j(f-v) . \tag{23}$$

Es existiert nun ein k mit $\|v-v^*\|_R = |L_k(v-v^*)|$.

Dann ist für dieses k:

$$\|f-v^*\| \geq \min_{1\leq i\leq m} L_i(f-v) + \|v-v^*\|_R \min\left(1, \frac{\lambda_k}{1-\lambda_k}\right)$$

$$\geq \min_{1\leq i\leq m} L_i(f-v) + \gamma\|v-v^*\|_R .$$

Beispiel (siehe Schaback [11]): Approximiert man $f(x,y) = x^2 + y^2$ bezüglich $V = \mathrm{span}(1,x,y)$ über $B = [-1,1]^2$, so ist $v^*(x) \equiv 1$ Minimallösung mit in Abbildung 1 dargestellter Extremalpunktmenge $\{t_1,...,t_5\}$

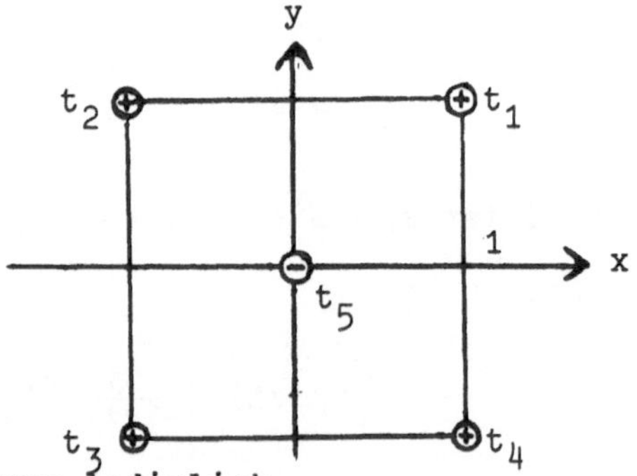

Abb. 1

Als minimale H-Mengen aus $E_n(f)$ kommen lediglich $R_1 = \{L_1, L_3, L_5\}$ und $R_2 = \{L_2, L_4, L_5\}$ in Frage mit

$$L_i(g) = g(t_i) \qquad \text{für } i = 1,...,4$$

und $$L_5(g) = -g(t_5) \qquad (g \in C(B)) .$$

Weder R_1 noch R_2 sind H_s-Mengen, da $\dim R_1 = \dim R_2 = 2$. Betrachtet man

$$R = R_1 \cup R_2$$

so ist R eine H_s-Menge und

$$\tfrac{1}{2}(L_1+L_3) + \tfrac{1}{4} L_5 \in V^{\perp} ,$$

$$\tfrac{1}{2}(L_2+L_4) + \tfrac{1}{4} L_5 \in V^{\perp} .$$

Somit folgt aus Satz 5:

$$\gamma = \tfrac{1}{3} .$$

Schaback [11] benutzt ebenfalls R als H_s-Menge, jedoch allein das Funktional

$$\tfrac{1}{8}(L_1+L_2+L_3+L_4) + \tfrac{1}{2} L_5 \in V^{\perp}$$

und gelangte somit nur zu $\gamma = 1/7$. Mit $\gamma = 1/3$ ist die Abschätzung scharf.

4. Konstruktion von H_s-Mengen.

Da die oben erwähnten Fehlerabschätzungen ihrem Wesen nach auf Dualitätsaussagen beruhen, wird man erwarten, daß gerade duale Algorithmen vom Remez-Typ in natürlicher Weise auf H_s-Mengen führen. Wir betrachten solche Algorithmen für unser eingangs genanntes Approximationsproblem (Carasso, Laurent [4], Ruffer-Beedgen [10]). All diesen Algorithmen ist folgendes gemeinsam: In jedem Iterationsschritt liegt eine <u>Referenzkette</u>

$$R = (\tilde{R}_1,\dots,\tilde{R}_s)$$

vor. Hierbei ist $\tilde{R}_1$ eine V-Referenz (≡ minimale H-Menge bezüglich V), $\tilde{R}_2$ eine V_1-Referenz mit $V_1 = V \cap [R_1]^\perp$, $\tilde{R}_3$ eine V_2-Referenz mit $V_2 = V_1 \cap [R_2]^\perp$ usw. (Dabei bezeichne $[R_1]$ den von R_1 in $V^\perp$ aufgespannten Unterraum). Außerdem hat man eine Minimallösung $\hat{v}$ dieser Kette R und einen Abweichungsvektor

$$h = (h_1, h_2,\dots,h_s) ,$$

der mit R folgendermaßen verknüpft ist:

Für alle $L_i \in \tilde{R}_i$ gilt

$$L_i(f-\hat{v}) = h_i \qquad (i = 1,\dots,s).$$

Betrachtet man nun

$$R = \bigcup_{i=1}^{s} \tilde{R}_i ,$$

so ist dim R = n und R ist eine H_s-Menge. Für jedes $L_i \in \tilde{R}_i$ kann man nämlich mit Hilfe des Austauschsatzes (Carasso, Laurent [4]) eine minimale H-Menge R_i mit $L_i \in R_i \subset R$ konstruieren, indem man die Referenz $\tilde{R}_i$ bis zur ersten Stelle der Referenzkette vortauscht. Das Ergebnis ist schließlich die gewünschte V-Referenz R_i mit $R_i \subset R$. Die λ_i ergeben sich als charakteristische Zahlen zu dieser Referenz.

Diese Vorgehensweise läßt sich anschaulich an dem vorigen Beispiel demonstrieren:

Dort haben wir zu der Minimallösung die Referenzkette

$$R = (R_1^{\sim}, R_2^{\sim})$$

mit $R_1^{\sim} = \{L_1, L_3, L_5\}$, $R_2^{\sim} = \{L_2, L_4\}$.

$R_1^{\sim}$ ist nämlich eine V-Referenz (= minimale H-Menge bezüglich V) und $R_2^{\sim}$ eine V_1-Referenz mit $V_1 = V \cap [R_1^{\sim}]^{\perp}$. Tauscht man nach dem Austauschsatz $R_2^{\sim}$ nach vorne, so erhält man gerade die neue V-Referenz

$$R_2 = \{L_2, L_4, L_5\}.$$

Man bekommt somit gerade die oben angegebenen H-Mengen.

Beispiel 2 (Curtis, Powell [6], Osborne, Watson [8]):
Die Funktion $f(x) = x^2$ soll über [0,2] bezüglich $V = \text{span}(x, e^x)$ approximiert werden. Die Minimallösung v* ist eindeutig,

$$v^*(x) = 0.18423256x + 0.41863122e^x$$

mit Minimalabweichung 0.53824532. Die Fehlerkurve hat jedoch nur 2 Extremas, nämlich bei dem Punkt 2 und bei $\xi = 0.40637574$ (alle Werte sind auf 8 geltende Ziffern genau [8]). Startet man den Remez Algorithmus (mit δ-Austausch und $\delta = 0.01$ [10]) bei den Punkten

$$0 \ , \ \eta = 0.406376 \ , \ 2 \ ,$$

so erhält man im zweiten Schritt eine zweistellige Referenzkette

$$R = (R_1, R_2)$$

mit $\quad R_1 = \{L_1, L_2\}$,

$$L_1(g) = g(\eta),$$

$$L_2(g) = -0.99999713\, g(2) + 0.00000287\, g(1)$$

($g \in C[0,2]$). In den Schritten 3 bis 9 operiert der Remez-Algorithmus nur noch auf der letzten Referenz in diesen Ketten. Jeweils R_1 bleibt unverändert. R_1 besteht im wesentlichen aus den Punktfunktionalen bei den endgültigen Extremalstellen 2 und ξ. Dies drückt sich in der zugehörigen Referenzabweichung

$$h_1 = 0.53824490$$

aus, die mit der Minimalabweichung auf 6 geltende Ziffern übereinstimmt. Im neunten Schritt hat man eine Referenzkette

$$R' = (R_1, R_2')$$

mit

$$R_2' = \{L_1', L_2'\} ,$$
$$L_1'(g) = g(0.40429688) ,$$
$$L_2'(g) = g(0.41210938) .$$

Im Algorithmus werden die vorliegenden Matrizen mittels Householdertransformationen auf "Dreiecksgestalt" reduziert. Diese Transformation liefert eine Basis $V = \text{span}\,\{v_1, v_2\}$ mit $[v_2] = [R_1]^\perp$, wie man sie in der Abschätzung (18) braucht:

$$v_1(x) = -0.96526597e^x - 0.26126918x ,$$
$$v_2(x) = -0.26126918e^x + 0.96526597x .$$

Bezüglich dieser Basis hat die Minimallösung die Form

$$v^* = \alpha\, v_1 + \beta\, v_2$$

mit $\alpha = -0.45222448$, $\beta = 0.06845805$.

Die im 9. Schritt berechnete Näherung lautet

$$\tilde{v} = \tilde{\alpha}\, v_1 + \tilde{\beta}\, v_2$$

mit $\tilde{\alpha} = -0.45222450$, $\tilde{\beta} = 0.07283348$, wobei sich $\tilde{\alpha}$ seit dem 2. Schritt nicht verändert hat. $\tilde{\alpha}$ stimmt mit α auf 7 geltende Ziffern überein. Aus (18) erhält man andererseits

$$\|\tilde{v}-v^*\|_{R_1} \leq 5\cdot 42\cdot 10^{-8} = 21\cdot 10^{-7} ,$$

denn $\min(\lambda_1,\lambda_2) \leq 0.168$. Daraus erhält man die Abschätzung für die Koeffizienten bezüglich v_1:

$$|\alpha-\tilde{\alpha}| < 3\cdot 10^{-7} .$$

Diese Abschätzung spiegelt recht gut die wirklichen Verhältnisse wieder.

Tauscht man R_2' an die erste Stelle der Referenzkette, so erhält man die neue V-Referenz

$$R_2^{\sim} = \{L_1', L_2', L_2\}$$

mit den zugehörigen charakteristischen Zahlen

$$0.123\ldots ;\ 0.045\ldots ;\ 0.831\ldots .$$

Wir erhalten dann aus (20):

$$|L_1'(\tilde{v}-v^*)| < 1826\cdot 10^{-7} .$$

Daraus ergibt sich wieder

$$|\beta-\tilde{\beta}| \leq 0.051 ,$$

eine Abschätzung, die die tatsächlichen Werte gut einschließt. Betrachtet man die Basis $\{x,e^x\}$ so sind im 9. Schritt die Koeffizienten der Näherung $\tilde{v}$ auf 2 geltende Ziffern genau. Die entsprechenden Abschätzungen geben auch dieses Verhalten wieder.

Literatur

1. Bartelt, M.W., McLaughlin: Characterizations of strong unicity in approximation theory, J. Approximation Theory 9 (1973), 255-266.
2. Bartelt, M.W., Schmidt, D.: On Poreda^s problem for strong unicity constants, preprint.
3. Blatt, H.-P., Klotz, V.: Zur Anzahl der Interpolationspunkte polynomialer Tschebyscheff-Approximationen im Einheitskreis, in Collatz,L., Meinardus, G., Werner, H. (Hrsg): Numerische Methoden der Approximationstheorie, ISNM 42, Birkhäuser, Basel 1978, 61-77.
4. Carasso, C., Laurent, P.J.: An algorithm of successive minimization in convex programming, R.A.I.R.O., Analyse numérique, Numerical Analysis (1978), 377-400.
5. Collatz, L.: Approximation von Funktionen bei einer und bei mehreren unabhängigen Veränderlichen, Z. Angew. Math. Mech. 36 (1956), 198-211.

6. Curtis, A.R., Powell, M.J.D.: Necessary conditions for a minimax approximation,Computer J. 8 (1966), 358-361.

7. Henry, M.S., Roulier, J.A.: Lipschitz and strong unicity constants for changing dimension, J. Approximation Theory 22 (1978), 85-94.

8. Osborne, M.R., Watson, G.A.: A note on singular minimax approximation problems, J. Math. Anal. Appl. 25 (1969), 692-700.

9. Poreda, S. J.: Counterexamples in best approximation, Proc. Amer. Math. Soc. 56 (1976), 167-171.

10. Ruffer-Beedgen, B.: Der Rémèz-Algorithmus ohne Haarsche Bedingung, Diplomarbeit, Universität Mannheim,1981.

11. Schaback, R.: Bemerkungen zur Fehlerabschätzung bei linearer Tschebyscheff-Approximation, in Collatz, L., Meinardus, G., Werner, H. (Hrsg): Numerische Methoden der Approximationstheorie, ISNM 52, Birkhäuser, Basel 1980, 255-276.

12. Schmidt, D.: On an unboundedness conjecture for strong unicity constants, J. Approximation Theory 24 (1978), 216-223.

Hans-Peter Blatt
Fakultät für Mathematik und Informatik
Universität Mannheim
A 5
D-6800 Mannheim

POLYNQM- UND SPLINE-INTERPOLATION

EIN FARBFILM

Klaus Böhmer

We describe a color movie about polynomial and spline interpolation available from the Institut für den wissenschaftlichen Film, Göttingen, W.-Germany.

Für das bekannte Beispiel von Runge [5] $f(x) = 1/(1+x^2)$ für $-5 \leq x \leq 5$ wird die Interpolation mit Polynomen und natürlichen kubischen Spline-Funktionen in äquidistanten Punkten und mit Polynomen in Tschebyscheff-Punkten gegenübergestellt. Dieser Farbfilm mit deutschem Tonkommentar ist beim Institut für den wissenschaftlichen Film, Göttingen, für Hochschulen und Schulen i. w. kostenlos ausleihbar.

Im ersten Teil des Films wird die Ausgangsfunktion f durch Polynome interpoliert. Da f eine gerade Funktion ist, wählen wir zu Interpolation 2n+1 äquidistante Punkte, die symmetrisch zur y-Achse liegen: $(x_i := 5(n-i)/n, f(x_i))$, $i=0,1,\ldots,2n$. Im Film werden Koordinatensystem, Ausgangsfunktion und die Interpolationspunkte grün und die Interpolationspolynome mit den Interpolationspunkten rot projiziert. Dadurch erscheinen die Interpolationspunkte und Bereiche guter Approximation in einer gelben Mischfarbe. Für wachsende 2n werden in Fig 1-3. einige dieser Polynome gezeigt.

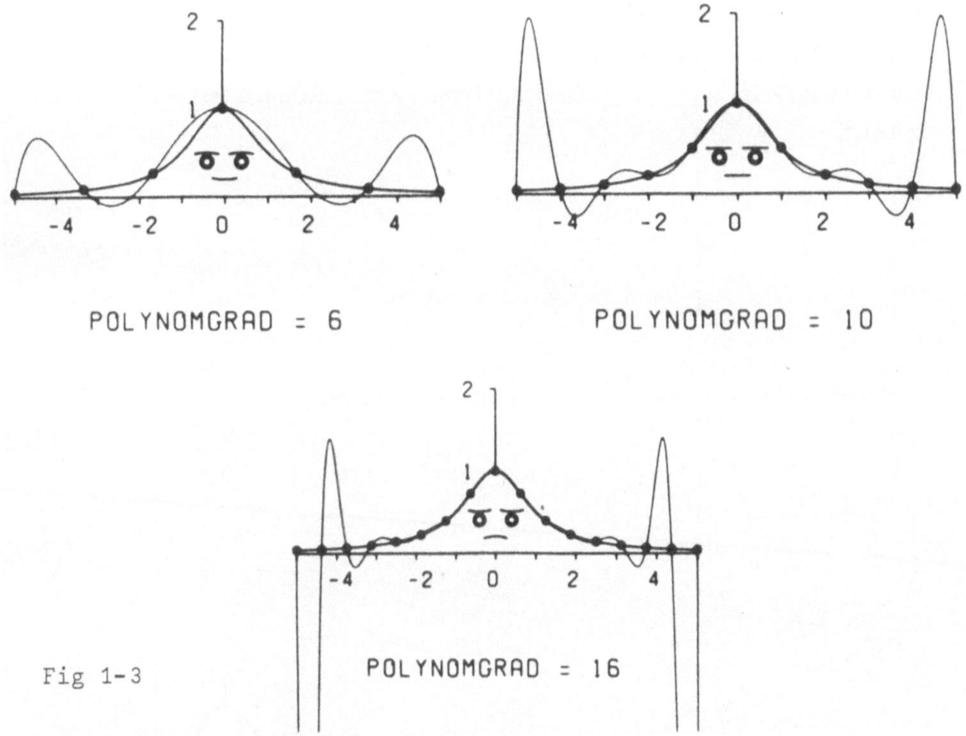

Fig 1-3

Es fällt auf, daß für wachsende 2n die Oszillationen in Randnähe immer unangenehmer werden und damit für die Divergenz der Folge von Interpolationspolynomen verantwortlich sind. Diese Oszillationen werden durch die Dynamisierung im Film noch weiter verdeutlicht. Für die nichtstatischen Übergänge zwischen Polynomen vom Grad **2n** und 2n+2, P_{2n} und P_{2n+2}, werden konvexe Linearkombinationen $(1-\lambda_\nu)P_{2n}+\lambda_\nu P_{2n+2}$, $0<\lambda_\nu<1$, gewählt. Die besonders interessanten Bereiche der Konvergenz (für $|x|\leq 3.63\ldots$) und der Divergenz (für $|x|\geq 3.6\ldots$) werden in eigenen Bildfolgen noch einmal projiziert.

Bedingt durch die schlechten Erfahrungen mit äquidistant interpolierenden Polynomen werden in der nächsten Szene äquidistant interpolierende natürliche kubische Splines herangezogen [1].

Wie zu erwarten, sind die Approximationseigenschaften sehr gut und die gelbe Mischfarbe (also optische Überdeckung der Ausgangs- und der Approximationsfunktion, und damit sehr gute Näherung) stellt sich schon ab 2n≥12 für das gesamte Intervall ein (Fig. 4-7). In diesen Figuren sind in der rechten Bildhälfte die Splinefunktionen, links die in Tschebyscheff-Punkten interpolierenden Polynome gezeigt.

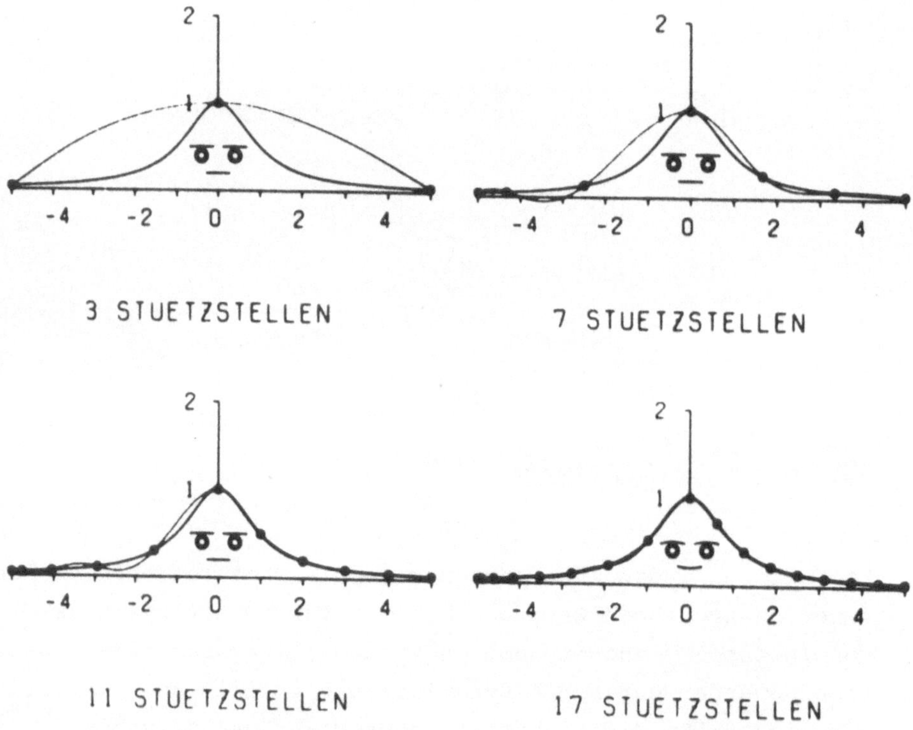

Fig 4-7

Da die Interpolation in äquidistanten Punkten die ungünstigste Situation für Polynome der günstigsten Situation für Splines gegenüberstellt, muß abschließend die "günstigste" Knotenwahl für Polynome zum Vergleich herangezogen werden, die Interpolation in Tschebyscheff-Knoten. Die minimale Supremumsnorm des Knotenpolynoms wird für die Tschebyscheffpunkte $\cos(\frac{2i+1}{4n+2}\pi)$, $i=0,1,\ldots,2n$ (bez. des Intervalls $[-1,1]$) angenommen [4]. "Fast optimale" Knoten im Sinne nahezu minimaler Norm des Projektionsoperators erhält man für die gestreckten Tschebyscheff-Punkte $(\cos\frac{2i+1}{4n+1}\pi)/\cos\frac{\pi}{4n+2}$, $i=0,1,\ldots,2n$ (vgl. [3] für Literaturangaben zu diesem Problem). Da wir wegen der Symmetrie und leichteren Vergleichsmöglichkeit den Knoten $x=0$ erzwingen wollen, wählen wir die Tschebyscheffpunkte $\cos\frac{i\pi}{2n}$, $i=0,\ldots,2n$, eine Art Zwischensystem zwischen den beiden anderen Systemen. Es zeigt sich, daß die in diesen Punkten interpolierenden Polynome, bis auf kleine Welligkeiten, sehr gut approximieren. Das sollte jedoch nicht über die Negativ-Resultate über Polynominterpolation hinwegtäuschen, wonach zu jeder Knotenfolge eine stetige Funktion existiert, deren Folge zugehöriger Interpolationspolynome divergiert.

Eine detaillierte Darstellung findet sich in der Begleitpublikation zum Film [2].

Literatur

1. Böhmer, K.: Spline-Funktionen, Teubner-Verlag, Stuttgart, 1974.

2. Böhmer, K., Kürschner, P.: Experimente mit Polynom- und Spline-Interpolation, Institut für den wissenschaftlichen Film. Begleitpublikation, Göttingen, 1981.

3. de Boor, C.: Polynomial Interpolation, Proceedings of the International Congress of Mathematicians, Helsinki, 917-922, 1978.

4. Davis, P.J.: Interpolation and Approximation, Blaisdell Publ. Comp., New York, 1963.

5. Runge, C.: Über empirische Funktionen und die Interpolation zwischen äquidistanten Ordinaten, Z. Math. u. Physik 46, (1901), 224-243.

Klaus Böhmer
Fachbereich Mathematik
Universität Marburg
Lahnberge
3550 Marburg
W.-Germany

A MULTIVARIATE ADAPTIVE DATA FITTING ALGORITHM

Michael Brannigan

A method is presented for solving the problem of approximating a discrete set of points in a multivariate setting using adaptive approximating families. The user has to make a choice of the approximating family, the norm and make assumptions about the distribution of errors. The criterion for selecting the best from the set of possible approximations is simple to apply and is derived from information-theoretic considerations. It is shown that under certain restrictive assumptions this method and that of generalized cross-validation are asymptotically equivalent. Numerical results are given to support the validity of the method.

INTRODUCTION

In view of the ever-increasing mass of data from diverse areas of science which require analysis, there is a need to develop automa= tic numerical data-fitting methods in a multivariate setting. What we propose in this paper is one approach to this problem, and we in no way present this as a universally acceptable method for all data sets: in fact we would caution the user of any method never blindly to ac= cept the resulting output but to attempt some experimental assessment.

Stated formally: given the data set (t_s, y_s), $s=1,\ldots,N$; $t_s \in \mathbf{R}^n$, $y_s \in \mathbf{R}$, we assume the existence of a relationship between t_s and y_s of the form

$$y_s = F(t_s) + e_s \quad , \quad s=1,\ldots,N,$$

where F is an unknown underlying function and e_s is the unknown error in the measurement of y_s. The problem that we address is: *find a good approximation to the function* F.

In summary our method takes an approximating family of functions with a variable number of parameters and a norm such that a best ap= proximation to the data can be found. For each best approximation the distribution of the residuals is compared, using an information-theoretic criterion; with a presumed form for the distribution of actual errors r_s, resulting in an assessment of the number of para= meters needed for the final approximation.

Other attempts at adaptively fitting functions to data are given in [5,8,9] with reports of success for various data sets. These methods, however, require assumptions concerning the data which we feel to be unnecessary and often inapplicable to general data sets. Thus in [5] uncorrelated errors are looked for to test for trends in the data, in [8] the analyst is asked to supply an upper bound for the errors in the data and in [9] normality of the distribution of errors is assumed. Our method, on the other hand, leaves to the user the choice of the approximating set of functions, the norm and the re= lated form for the distribution of the errors.

METHOD

(a) Approximating set

The first decision to be made is which set of approximants should be used for the given data set. If any information concerning the data is available to the analyst, then the choice should of course reflect this knowledge. In most cases families of functions are chosen where the number of parameters can vary, such as splines.

DEFINITION. *An approximating family of functions*

$$S_k = \{\phi(t,\theta_k) : t \in \mathbf{R}^n \quad , \quad \theta_k \in \mathbf{R}^k\}$$

is adaptive if S_k *is defined for all* k *such that* $1 \le k \le L$, *for some* L.

In practice piecewise polynomials in some form represent the most useful set for general programs. Adaptivity in this setting is the ability to increase the number of knots. The reason for this choice is that most continuous functions arising in practice can be well approximated by such a family of approximants using small numbers of knots and hence a small number of parameters.

(b) Norm

The question as to which norm is to be used is crucial, as our numerical experiments have shown. One guideline which is well founded is the following: if the distribution of errors has the form:

$$\exp(-|x|^p/p\,\sigma^p)/\sigma$$

then use of the ℓ_p norm provides maximum likelihood estimates of our parameters θ_k in the case when ϕ depends linearly on θ_k. For a discussion of which norm to use for smoothing, see [6].

The necessity of choosing an approximating set and norm is fundamental in data-fitting methods, and if the information available about the underlying function F is sufficient to enable the value of k from our approximating family to be known, then the pro= blem is well-defined and we can obtain θ_k^* such that

$$\| y_s - \phi(t_s,\theta_k^*) \| = \min_{\theta_k \in \mathbf{R}^k} \| y_s - \phi(t_s,\theta_k) \| ,$$

which gives us the required result.

In the general setting of no information we can obtain $\theta_1^*,\theta_2^*,\ldots,\theta_L^*$ and we are left with the problem of deciding which one of the θ_i^* is best. We must assume that our choice of ϕ and norm is such that for some m

$$\| \phi(t_s,\theta_m^*) - F(t_s) \|$$

is small. Our problem has now been reduced to the question: *what is the value of* m?

To solve this problem we have to make one further assumption. As our analysis so far has not taken into consideration the errors e_s and the residuals of our approximations, we must consider these values.

(c) Information criterion

For any $\theta_k \in \mathbf{R}^k$ we have

$$y_s = \phi(t_s,\theta_k) + r_s(\theta_k) \quad , \quad s=1,\ldots,N.$$

The residuals $r_s(\theta_k)$ of the approximation can be computed and we should compare these known residuals with the unknown errors e_s of the data. The minimum amount of information which we can prescribe to these errors is the form of their distribution.

Let $f(x,\xi)$ be the assumed distribution for the errors, where $x \in \mathbf{R}$ is a random variable and ξ the vector of parameters. For each θ_k the residuals $r_s(\theta_k)$ of the approximation are considered as N ob= servations of the random variable x; the vector ξ is then calculated as the maximum likelihood estimate, given $\theta_k \in \mathbf{R}^k$. Our assumption concerning this choice of distribution is that for some $\Theta \in \mathbf{R}^m$ the resi= duals $r_s(\Theta)$ equal e_s, s=1,...,N; the estimate $\xi(\Theta)$ we shall denote by Ξ. It is our intention, therefore, to compare $f(x,\xi(\theta_k))$ with

$f(x,\Xi)$, and to do this we use the mean information for discrimination as defined in [3, p. 5], viz.

$$I(\Xi\,,\xi(\theta_k)) = \int f(x\,,\Xi)\,\ell n\,\frac{f(x\,,\Xi)}{f(x\,,\xi(\theta_k))}\,dx\,.$$

We then choose that k which minimizes the expected value of I; this is related to the maximum entropy principle, where I is the entropy.

Using the same analysis as [3, p. 27] we obtain for a small per= turbation $\Delta\Theta$ from Θ

$$I(\Xi\,,\xi(\Theta + \Delta\Theta)) = \tfrac{1}{2}\,\Delta\Theta\, J\, \Delta\Theta^T \qquad \text{(T denotes transpose)}$$

$$= \tfrac{1}{2}\,\|\Delta\Theta\|_J^2$$

where J is the Fisher information matrix defined by

$$J_{ij} = -\int f(x\,,\Xi)\,\frac{\partial^2\,\ell n\, f(x\,,\Xi)}{\partial\theta_i\partial\theta_j}\,dx\,.$$

If we now consider elements of $\mathbf{R}^k$ as elements in $\mathbf{R}^m$ with the compo= nents k+1,...,m equal to zero, we are able to write for θ_k^* (the best approximation vector in $\mathbf{R}^k$) and θ_k the best estimate of Θ in $\mathbf{R}^k$:

$$2\,I\,(\Xi\,,\xi(\theta_k^*)) = \|\theta_k^* - \Theta\|_J^2$$

$$= \|\theta_k^* - \theta_k\|_J^2 + \|\theta_k - \Theta\|_J^2 + 2(\theta_k^* - \theta_k)J(\theta_k - \Theta)^T.$$

From Taylor's theorem

$$\ell n\; f(x\,,\xi(\theta_k)) = \ell n\; f(x\,,\xi(\theta_m^*)) + \tfrac{1}{2}(\theta_k - \theta_m^*)J_1(\theta_k - \theta_m^*)^T$$

$$= \ell n\; f(x\,,\xi(\theta_k^*)) + \tfrac{1}{2}(\theta_k - \theta_k^*)J_2(\theta_k - \theta_k^*)^T\,,$$

where

$$(J_1)_{ij} = \frac{\partial^2\,\ell n\, f(x\,,\xi(\theta_m^* + \rho_1(\theta_k - \theta_m^*))}{\partial\theta_i\partial\theta_j}$$

and

$$(J_2)_{ij} = \frac{\partial^2 \ell n\, f(x\,,\xi(\theta_k^* + \rho_2(\theta_k - \theta_k^*))}{\partial\theta_i \partial\theta_j}$$

for some $0 \le \rho_1, \rho_2 \le 1$.

Hence for any constant K we have

$$\ell n \frac{f(x\,,\xi(\theta_k^*))}{f(x\,,\xi(\theta_m^*))} = \frac{K}{2} f(x\,,\Xi)\big((\theta_k-\theta_m^*)J_1(\theta_k-\theta_m^*)^T - (\theta_k-\theta_k^*)J_2(\theta_k-\theta_k^*)^T\big).$$

For N observations x_i, i=1,...,N let $K = (x_N - x_1)$, then

$$\sum_{i=1}^{N} \frac{K}{N} f(x_i\,,\Xi)J_p\Big|_{x_i} \rightarrow -J \quad , \quad p=1,2$$

where $J_p\big|_{x_i}$ is the matrix J_p evaluated at x_i.

Hence we have for the statistic

$$\begin{aligned}\nu(k,m) &= -2\sum \ell n \frac{f(x_i\,,\xi(\theta_k^*))}{f(x_i\,,\xi(\theta_m^*))} \\ &= N\,\|\theta_k - \theta_m^*\|_J^2 - N\|\theta_k - \theta_k^*\|_J^2 \\ &= N\,\|\theta_k - \Theta\|_J^2 + N\,\|\Theta - \theta_m^*\|_J^2 - N\,\|\theta_k - \theta_k^*\|_J^2 \\ &\qquad + 2N(\theta_k - \Theta)J(\Theta - \theta_m^*)^T.\end{aligned}$$

The random variable $N\,\|\Theta - \theta_m^*\|_J^2 - N\,\|\theta_k - \theta_k^*\|^2$ is χ^2-distributed with m-k degrees of freedom, and for k=m, $\theta_k - \theta_k^* = \Theta - \theta_m^*$; thus an estimate for

$$N\,\|\theta_k - \Theta\|_J^2 + 2N(\theta_k^* - \theta_k)J(\theta_k - \Theta)^T$$

is given by

$$(\nu(k,m) + k - m).$$

Also $N \| \theta_k^* - \theta_k \|_J^2$ is χ^2-distributed with k degrees of freedom. For a similar analysis using maximum-likelihood estimates for θ_k, see [4,10].

Hence an estimate for the expected value of $2\,I(\Xi\, , \xi(\theta_k^*))$ is given by

$$(\nu(k,m) + 2k - m)/N.$$

As was pointed out above, we wish to find that k which minimizes this expected value; hence, deleting constants in this expression, we arrive at the following result:

choose that k *which minimizes*

$$-2 \sum_{i=1}^{N} \ell n\, f(r_i(\theta_k^*)\, , \xi(\theta_k^*)) + 2k.$$

This result is similar to that given in [1], where maximum-likelihood estimates for θ_k are used throughout.

COMPARISON WITH CROSS-VALIDATION

When the function ϕ depends linearly on θ_k, the method described above may be considered as a problem in best subset regression, which can be solved using the ideas of generalized cross-validation under certain assumptions.

Assume that e_s, s=1,...,N are distributed normally as $N(0\, , \sigma^2)$, and that the approximation by $\phi(t\, , \theta)$ is done using the least-squares norm. Let X(k) be the design matrix set up for the approximation with k parameters; we then have

$$\theta_k^* = (X^T(k)X(k))^{-1}X^T(k)y.$$

Defining

$$A(k) = X(k)(X^T(k)X(k))^{-1}X^T(k)$$

the cross-validation method would choose k such that v(k), given by

$$v(k) = \frac{1}{N} \| (I - A(k))y \|^2 / [\frac{1}{N} \mathrm{Trace}(I - A(k))]^2$$

is minimized.

If the singular-value decomposition of X(k) is

$$U \Sigma V^T,$$

then

$$A(k) = U \Sigma (\Sigma \Sigma^T)^+ \Sigma U^T$$

and

$$\mathrm{Trace}\ (A(k)) = k.$$

In the method given above we wish to find that k which minimizes

$$w(k) = N \ell n(\| (I - A(k))y \|^2 /N) + 2k.$$

From $(w(k) - 2k)/N = \ell n(\| (I - A(k))y \|^2 /N)$ we have

$$\exp(w(k)/N) = \frac{\| (I - A(k))y \|^2 /N}{\exp(- 2k/N)}$$

$$\rightarrow \frac{\| (I - A(k))y \|^2 /N}{(1 - k/N)^2} \quad \text{as} \quad \frac{k}{N} \rightarrow 0$$

$$= v(k).$$

Hence under these assumptions generalized cross-validation is asymptotically equivalent to our method. For a comparison between ordinary cross-validation and the methods of [1], see [7].

The assumptions under which cross-validation holds are very restrictive; they are linearity, normality and least-squares approximation. The method that we describe covers non-linearities, non-normality and allows any appropriate norm for approximation.

NUMERICAL EXAMPLES

Our first experiment consisted of fitting Hermite-cubic poly= nomials to 200 points taken from the function $x^{2/3}$ within the range [-1 , 1], a function that exhibits a discontinuity in the derivative at $x = 0$. Fig. 1 shows the result of our method when the points were contaminated by a uniform distribution of errors within the range [-0.2 , 0.2] and the approximation was carried through using the mini= max norm. The same set of points, but contaminated by errors from a normal $N(0, 0.1)$ distribution, were approximated using the ℓ_2 norm, and Fig. 2 shows the result.

To emphasize how crucial the choice of norm is to the method, we took 142 data points from an experiment determining the angular dis= placement of a horizontal elbow flexion against time, and contami= nated these data points with random errors taken from a Cauchy dis= tribution. Assuming normality of the underlying distribution, Fig. 3 shows the result using ℓ_1 approximation and cubic splines as approxi= mants. This result agreed well with the original data, regardless of the 'wild' points. Using the minimax norm the result, shown in Fig. 4, differed dramatically. In all our numerical experiments we found that choice of norm and not of distribution was the most crucial decision. For further results and a discussion of a numerical method in the single variable case, see [2].

Theoretically our method is valid for multivariable problems. In the multivariate setting, however, the choice of ϕ is difficult and the obvious choice of tensor product splines is not always the best. For uniformly spread data points and a smooth underlying func= tion, tensor product splines are effective, as we can see in Fig. 5. Here we have the underlying function $F = (2x^2 - 1)y$ represented by 200 points on the square $[-1, 1] \times [-1, 1]$, contaminated by errors from a normal distribution. Fig. 5 is the result of using bicubic splines, least-squares norm and normal distribution.

Further numerical work needs to be done in the multivariate set= ting, especially regarding the choice of approximating families, but our method does yield a theoretical basis for adaptively fitting such approximants to a large variety of data sets.

REFERENCES

[1] Akaike, H.: Information Theory and an Extension of the Maximum Likelihood principle. Proc. 2nd Symp. Inf. Th. (B.N. Petrov & F. Csaki Eds.) Akademie Kiado, Budapest 1973, 267-281.

[2] Brannigan, M.: An Adaptive Piecewise Polynomial Curve Fitting Procedure for Data Analysis. Comm. Stats. A10, 18 (1981).

[3] Kullback, S.: Information Theory and Statistics. New York, John Wiley 1959.

[4] Kupperman, M.: Probabilities of Hypotheses and Information Statistics in Sampling from Exponential-Class Populations. Ann. Math. Stats. 29 (1958), 571-574.

[5] Powell, M.J.D.: Curve Fitting by Splines in One Variable. Numerical Approx. to Functions and Data (J.G. Hayes Ed.) Athlone Press 1970, 65-83.

[6] Rice, J.R. & White, J.S.: Norms for Smoothing and Estimation. SIAM Review, 6 (1964), 243-256.

[7] Stone, M.: An Asymptotic Equivalence of Choice of Model by Cross-Validation and Akaikes Criterion. J. Roy. Stat. Soc., B36 (1977), 44-47.

[8] Taylor, G.D. & Vogel, C.R.: An Adaptive Multidimensional Data Fitting Package. Private communication

[9] Wahba, G. & Wold, S.: A Completely Automatic French Curve: Fitting Spline Functions by Cross-Validation. Comm. Statistics, 4 (1975), 1-17.

[10] Wald, A.: Tests of Statistical Hypotheses Concerning Several Parameters when the Number of Observations is Large. Trans. Am. Math. Soc. 54 (1943), 426-482.

Prof. Dr. M. Brannigan
Visiting Lecturer
The University of Georgia
Department of Statistics and
Computer Science
Graduate Studies Building
Athens, Georgia 30602, USA

MdeV

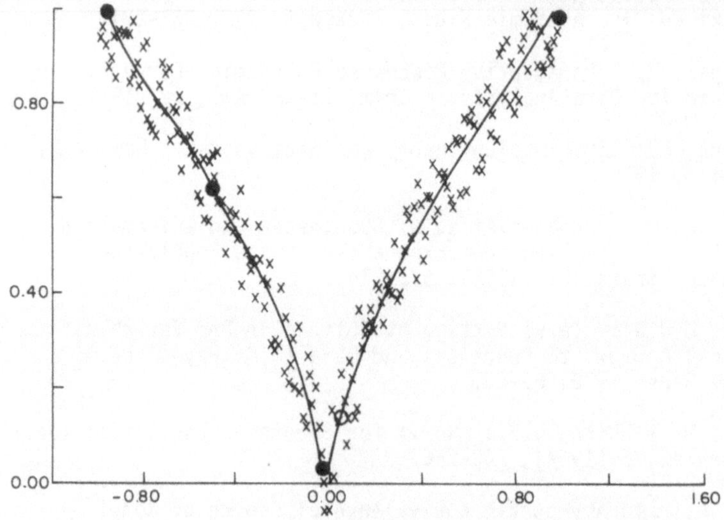

FIGURE 1

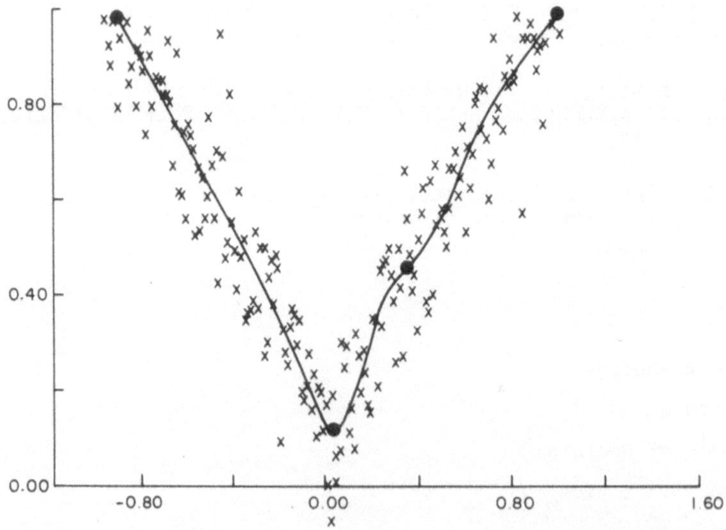

FIGURE 2

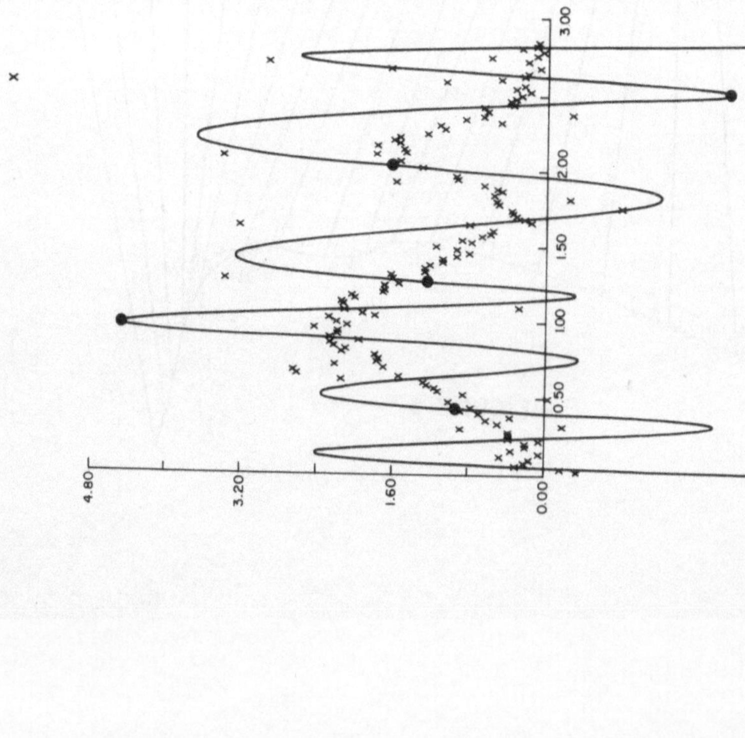

FIGURE 4

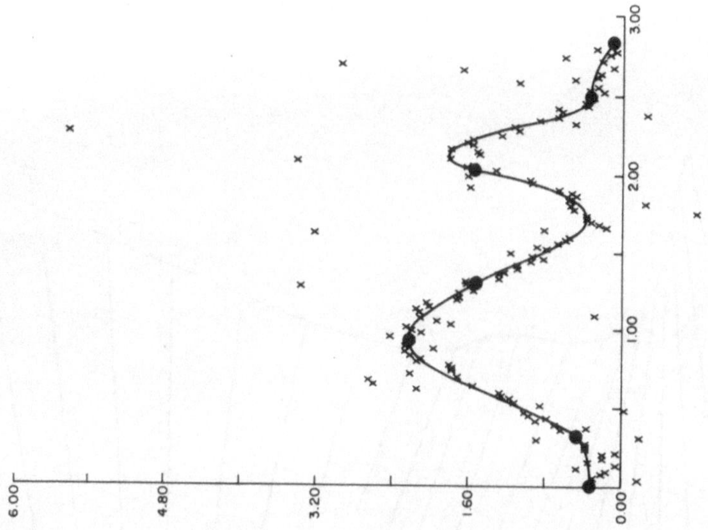

FIGURE 3

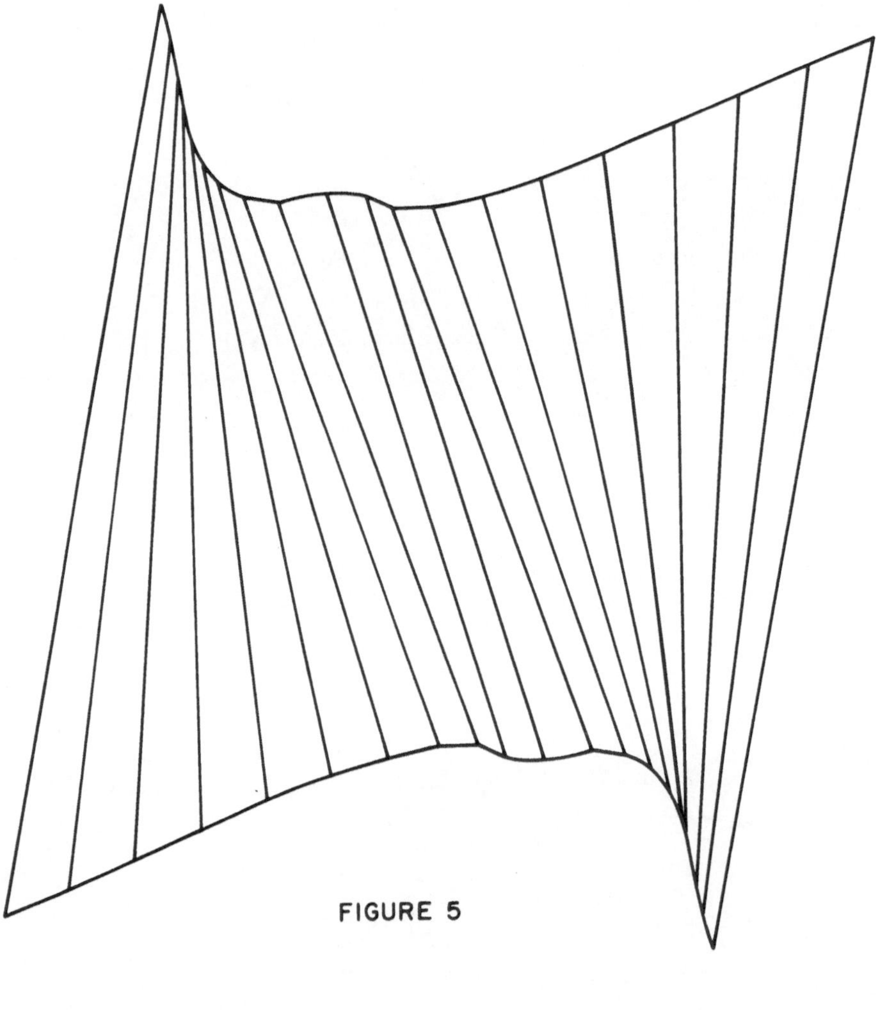

FIGURE 5

ZUR NUMERISCHEN BERECHNUNG KONJUGIERTER FUNKTIONEN

Helmut Braß

The asymptotic behaviour of the optimal bounds for the error which occurs in the numerical calculation of conjugate functions by Wittichs method is determined. A modification of the Wittich algorithm reduces the error bounds, but the order of the error is unimprovable by any method based on the same information.

1. Übersicht

Im folgenden sei f eine (2π)-periodische Funktion mit beschränkter κ - ter $(\kappa \geq 1)$ Ableitung. Ihre Fouriersche Reihe sei

(1) $$f(x) = \sum_{\nu=-\infty}^{\infty} \gamma_\nu e^{i\nu x} .$$

Die zu f "konjugierte Funktion" K[f] wird definiert durch

$$K[f](x) = -i \sum_{\nu=-\infty}^{\infty} \gamma_\nu e^{i\nu x} \operatorname{sgn}\nu .$$

Die praktische Aufgabe der Berechnung konjugierter Funktionen besteht darin, aus den vorgegebenen Werten von f an den Stellen

(2) $$x_j = j n^{-1} \qquad (j=-n+1,\ldots,n)$$

Schätzwerte für K[f] entweder an allen Stellen x oder nur an den Stellen x_j herzuleiten. Diese Aufgabe ist besonders wichtig in der praktischen konformen Abbildung (siehe Gaier [1]), und zahlreiche Algorithmen sind für sie angegeben worden (Gaier [1] S. 74 ff, Gutknecht [2], Knauff/Kreß [4], Onegov [7]). Unter diesen wird der Algorithmus von Wittich wohl am häufigsten angewandt. Er besteht darin, K[f] näherungsweise zu ersetzen durch $K[\mathrm{intpol}_n[f]]$, hier bedeutet $\mathrm{intpol}_n[f]$ den trigonometrischen Interpolationsausdruck n - ter Ordnung zu den Stützstellen (2). Der Fehler

$$R_n^{Wi}[f] = K[f] - K[\mathrm{intpol}_n[f]]$$

soll hier diskutiert werden. Gaier ([1] S. 77) hat

$$R_n^{Wi}[f] = O(n^{-\kappa} \ln^2 n)$$

gezeigt, diese Größenordnung wird hier zu der nun unverbesserbaren Ordnung $O(n^{-\kappa} \ln n)$ verschärft werden.

Zur Formulierung des Resultates wird benötigt der Euler-Spline E_r

$$E_r(x) = 2 \sum_{\nu=-\infty}^{\infty} \frac{e^{i(2\nu+1)\pi x}}{[i(2\nu+1)\pi]^{r+1}} \qquad r=0,1,\ldots$$

und der Bernoulli-Monospline B_r

$$B_r(x) = - \sum_{\substack{\nu=-\infty \\ \nu=0}}^{\infty} \frac{e^{i\,2\nu\pi x}}{[i\,2\nu\pi]^r} \qquad r=1,2,\ldots$$

sowie die Vereinbarung

$$||f|| = \sup_x |f(x)| \quad .$$

Damit gilt

Satz 1 $$\sup_{\|f^{(\kappa)}\| \le 1} |R_n^{Wi}[f](x)| = a(x,\kappa,n)n^{-\kappa}\ln n + O(n^{-\kappa})$$

(der O- Term gilt gleichmäßig in x) mit

(3) $$a(x,\kappa,n) = \pi^{\kappa-1} \int_0^1 |E_{\kappa-1}(z)\cos nx + 2B_\kappa(z)|dz$$

und

(4) $$a_\kappa = a\left(\frac{\pi}{2n},\kappa,n\right) \le a(x,\kappa,n) \le a(0,\kappa,n) = 2^\kappa a_\kappa$$

mit

(5) $$a_\kappa = 2\pi^{\kappa-1} \int_0^1 |B_\kappa(z)|dz$$

Bemerkung 1 : $f^{(\kappa)}$ werde hier stets im verallgemeinerten Sinne verstanden (g ist verallgemeinerte κ-te Ableitung von f, wenn die gewöhnliche Ableitung $f^{(\kappa-1)}$ sich als Integral von g schreiben läßt).

Bemerkung 2 : Aus der Translationsinvarianz von R_n^{Wi} folgt, daß $\sup_{\|f^{(\kappa)}\| \le 1} |R_n^{Wi}[f](\cdot)|$ und damit $a(\cdot,\kappa,n)$ die Periode πn^{-1} hat, also insbesondere $a(x_j,\kappa,n) = a(0,\kappa,n)$ gilt. Wegen (4) bedeutet das, daß gerade an den besonders interessierenden Stellen x_j der (Maximal-)Fehler groß ist, in der Mitte zwischen zwei Stützstellen ist er merklich kleiner.

Bemerkung 3 : Restringiert auf [0,1] sind E_κ und B_κ bekanntlich Polynome κ-ten Grades, die Auswertung von (3) bereitet also keine prinzipiellen Schwierigkeiten.

Es ist

$$a_1 = \frac{1}{2} \ , \quad a_2 = \frac{\pi\sqrt{3}}{27} \ , \quad a_3 = \frac{\pi^2}{96} \ ,$$

statt weiterer Zahlen genüge die Angabe, daß $2^\kappa a_\kappa$ für alle $\kappa > 1$ nahe bei $8\pi^{-2}$ liegt.

Bemerkung 4 : Aus Satz 1 kann man mit Standard-Methoden noch folgern

$$\text{(6)} \qquad \|R_n^{Wi}[f]\| \le \text{const}_\kappa \frac{\ln n}{n^\kappa} \, \omega(f^{(\kappa)}; \frac{\pi}{n})$$

(ω der Stetigkeitsmodul). Denn nach Konstruktion gehören die trigonometrischen Ausdrücke (n-1)-ter Ordnung zum Kern von R_n^{Wi}, also gilt für jeden solchen Ausdruck t noch Satz 1

$$\|R_n^{Wi}[f]\| = \|R_n^{Wi}[f-t]\| \le \text{const}_\kappa \frac{\ln n}{n^\kappa} \|f^{(\kappa)} - t^{(\kappa)}\| \ ,$$

und man erhält (6) durch passende Wahl von $t^{(\kappa)}$. Dazu reicht der Jacksonsche Satz der Approximationstheorie aber nicht ganz aus, weil $t^{(\kappa)}$ kein absolutes Glied haben darf. Jedoch hat auch $f^{(\kappa)}$ den Mittelwert Null, man überzeugt sich leicht, daß die üblichen Beweise des Jacksonschen Satzes (siehe z.B. Meinardus [5]) die hier notwendige Verschärfung liefern.

Ohne Beweis sei noch hinzugefügt, daß (6) für $\kappa = 0$ auch dann nicht allgemein richtig ist, wenn man sich nur auf Funktionen f beschränkt, für die K[f] als stetige Funktion existiert.

Es ist bekannt, daß

$$\text{intpol}_n[f](x) = \sum_{\nu=0}^{n} \alpha_\nu^{(n)} \cos\nu x + \beta_\nu^{(n)} \sin\nu x \qquad (\beta_n^{(n)} = 0)$$

die Funktion f asymptotisch mit einem Fehler $O(n^{-\kappa} \ln n)$ annähert, und daß diese Approximationsgüte zu $O(n^{-\kappa})$ verbessert werden kann, wenn man die Interpolationspolynome in geeigneter Weise modifiziert. Eine solche Modifikation sind die "typical means" (Zygmund [8]) der Ordnung $\tau(\geq\kappa+1)$:

$$\mathrm{intpol}_n^{tm\sigma}[f](x) = \sum_{\nu=0}^{n} [1-(\tfrac{\nu}{n})^{\sigma}](\alpha_\nu^{(n)}\cos\nu x+\beta_\nu^{(n)}\sin\nu x) \quad ,$$

eine andere (nur für $\kappa=1,2$) die Bernstein-Rogosinski-Mittel:

$$\mathrm{intpol}_n^{BR}[f](x) = \sum_{\nu=0}^{n} \cos\nu \frac{\pi}{2n} (\alpha_\nu^{(n)}\cos\nu x+\beta_\nu^{(n)}\sin\nu x).$$

Es liegt hiernach nahe, das Wittich-Verfahren entsprechend zu modifizieren, d.h. nicht $\mathrm{intpol}_n[f]$ sondern $\mathrm{intpol}_n^{tm\sigma}[f]$ zu konjugieren. Für die algorithmische Brauchbarkeit ist wichtig, daß ein solches Vorgehen nicht wesentlich rechenaufwendiger ist als das Wittichsche, insbesondere kann die schnelle Fourier-transformation herangezogen werden (Gutknecht [2]). Es gilt

Satz 2 Sei

$$R_n^{tm\sigma}[f]= K[f] - K\,[\mathrm{intpol}_n^{tm\sigma}[f]].$$

Dann ist, sofern $\sigma \geq \kappa+1$

$$\sup_{\|f^{(\kappa)}\|\leq 1} |R_n^{tm\sigma}[f](x)|= a_\kappa \frac{\ln n}{n^\kappa} + O(n^{-\kappa}),$$

wobei a_κ in (5) definiert ist.

Man gewinnt also an den Stützstellen x_i an Genauigkeit, jedoch ist die Fehlergrößenordnung nicht geändert. Das läßt sich auch nicht erreichen, wenn man beliebige andere Konjugations-Algorithmen heranzieht, es gilt nämlich

Satz 3 Es gibt eine nur von κ abhängende Zahl d_κ derart, daß für jede reelle Funktion $\tilde{K}$ von 2n Veränderlichen

$$\sup_{\| f^{(\kappa)} \| \leq 1} |K[f]\,(0) - \tilde{K}(f(x_{-n+1}),\ldots,f(x_n))| \geq d_\kappa \frac{\ln n}{n^\kappa}$$

gilt.

Die zum Beweis der Sätze 1,2 herangezogene Methode erlaubt den Beweis weiterer Sätze aus dem Umkreis der trigonometrischen Interpolation, so zum Beispiel

Satz 4 Ist $\kappa > \Theta$, so gilt

$$\sup_{\| f^{(\kappa)} \| \leq 1} |f^{(\Theta)}(x) - \text{intpol}_n[f]^{(\Theta)}(x)| =$$

$$\phi_n(x)\ \pi^{\kappa-1}\ n^{-\kappa+\Theta} \ln n \int_0^1 |E_{\kappa-1}(z)|\,dz + O(n^{-\kappa+\Theta})$$

mit
$$\phi_n(x) = \begin{cases} \cos nx & \Theta \text{ ungerade} \\ \sin nx & \Theta \text{ gerade} \end{cases}$$

Der Fall $\Theta = 0$ dieses Satzes ist von Nikolsky [6] mit einer wesentlich anderen Methode bewiesen worden. Zur Differentiation der trigonometrischen Interpolationsausdrücke vergleiche man auch Haverkamp [3].

Schließlich sei noch darauf hingewiesen, daß auch die Approximationsgüte der Ableitungen der konjugierten Interpolationsausdrücke sich ganz ähnlich abschätzen läßt.

2. Das grundlegende Lemma

Ausgehend von (1) erkennt man, daß das asymptotische Verhalten von

(7) $$R_n[f] = \sum_{\nu=-\infty}^{\infty} \gamma_\nu \rho_{\nu,n}$$

zu diskutieren sein wird, dabei sind $\rho_{\nu,n}$ die Fehler des jeweils betrachteten Verfahrens angewandt auf $e^{i\nu\cdot}$. Die Vertauschung des Restoperators und der Summation ist in allen hier betrachteten Fällen leicht zu rechtfertigen. Es ist

$$\text{intpol}_n[e^{i\nu\cdot}](x) = \begin{cases} e^{i\mu x} & \mu \neq n \\ \frac{1}{2}e^{inx}+\frac{1}{2}e^{-inx} & \mu = n \end{cases}$$

wobei stets μ definiert wird durch

$$\nu = 2\lambda n + \mu \qquad (-n<\mu\leq n).$$

Demnach ist

(8) $$\rho_{\nu,n} = \begin{cases} b_\nu\, e^{i\nu x} - c_{\mu,n}\, e^{i\mu x} & \mu\neq n \\ b_\nu\, e^{i\nu x} - \frac{1}{2}c_{n,n}e^{inx} - \frac{1}{2}\, c_{-n,n}e^{-inx} & \mu=n \end{cases}$$

(9) $$\text{mit } b_{-\nu} = \bar{b}_\nu \quad \text{und} \quad c_{-\mu,n} = \bar{c}_{\mu,n}$$

eine Form für die $\rho_{\nu,n}$, die alle hier betrachteten Fälle erfaßt, nämlich das Wittich-Verfahren für

$$b_o = c_{o,n} = 0 \qquad b_\nu = c_{\nu,n} = i \qquad (\nu=1,\dots),$$

das modifizierte Wittich-Verfahren aus Satz 2 durch

$$b_o = c_{o,n} = 0\ ,\quad b_\nu = i\ ,\quad c_{\nu,n} = i\ (1-(\tfrac{\nu}{n})^\tau) \qquad (\nu=1,\ldots),$$

die Situation von Satz 4 durch

$$b_\nu = c_{\nu,n} = (i\nu)^\Theta \qquad (\nu=0,1,\ldots).$$

Weiter sei noch vorausgeschickt, daß es im Hinblick auf unser Ziel genügt, nur x-Werte mit

$$(10) \qquad |x| \leq \pi n^{-1}$$

zu betrachten, siehe Bemerkung 2 (erste Hälfte) zu Satz 1, die sinngemäß auch für Satz 2 und Satz 4 gilt.

Lemma 1. Es sei b_ν für $\nu \geq 1$ ein Polynom in ν vom Grad $\Theta < \kappa$ und ferner sei

$$\sup_{\mu\geq 1} |c_{\mu,n}| = O(n^\Theta)$$

$$\sup_{\mu\geq 1} |c_{\mu+1,n} - c_{\mu,n}| = O(n^{\Theta-1})$$

$$\sup_{\mu>1} |c_{\mu+2,n} - 2c_{\mu+1,n} + c_{\mu,n}| = O(n^{\Theta-2})$$

$$c_{o,n} = b_o$$

Dann gilt mit den durch (8) und (9) definierten $\rho_{\nu,n} = \rho_\nu$ unter der Voraussetzung (10)

$$R_n[f] = \frac{1}{4\pi}\ (\frac{-\pi}{n})^\kappa \int_{I_2} f^{(\kappa)}(y)\ \mathrm{ctg}\ \frac{y}{2}\ \{2\mathrm{Im}(c_1)B_\kappa(\frac{ny}{\pi}) +$$

$$\mathrm{Im}(c_n e^{inx})E_{\kappa-1}(\frac{ny}{\pi})\}dy$$

$$- \frac{(-i)^\kappa}{8\pi} \int_{I_2} f^{(\kappa)}(y) \sin^{-2}\frac{y}{2} \{ \sum_{|\nu|<n-1} e^{-i\nu y} [(\nu+1)^{-\kappa}\rho_{\nu+1} - 2\nu^{-\kappa}\rho_\nu$$

$$+ (\nu-1)^{-\kappa}\rho_{\nu-1}]dy + O(\{n^{-1}\sum_{\nu=1}^{n-1} |b_\nu - c_{\nu,n}|^2\}^{1/2})$$

$$+ O(n^{\Theta-\kappa})$$

Hier ist $I_2 = [x_{-n}, x_{-1}] \cup [x_1, x_n]$.

Beweis: Für alle $|\nu| \geq n-1$ gilt

(11) $\quad |\rho_\nu| = O(\nu^\Theta)$

wie ganz leicht zu sehen; und weiter gilt, sofern die vorkommenden Indizes $\not\equiv 0 \pmod n$ sind und $|\nu| \geq n-1$ ist

(12) $\quad |\rho_{\nu-1} - \rho_\nu| = O(n^{-1}\nu^\Theta)$

(13) $\quad |\rho_{\nu-1} - 2\rho_\nu + \rho_{\nu+1}| = O(n^{-2}\nu^\Theta)$.

Um etwa (12) zu beweisen, geht man so vor:

$$|\rho_{\nu-1} - \rho_\nu| = |b_{\nu-1}\, e^{i(\nu-1)x} - b_\nu e^{i\nu x} - c_{\mu-1} e^{i(\mu-1)x} + c_\mu e^{i\mu x}|$$

$$= |(b_{\nu-1} - b_\nu) e^{i(\nu-1)x} + b_\nu (e^{i(\nu-1)x} - e^{i\nu x}) - (c_{\mu-1} - c_\mu) e^{i(\mu-1)x}$$

$$+ c_\mu (e^{i(\mu-1)x} - e^{i\mu x})| \leq |b_{\nu-1} - b_\nu| + |b_\nu| |2 \sin\frac{x}{2}|$$

$$+ |c_{\mu-1} - c_\mu| + |c_\mu|\, |2 \sin\frac{x}{2}| = O(\nu^{\Theta-1}) + O(\nu^\Theta n^{-1})$$

$$+ O(n^{\Theta-1}) + O(n^\Theta\, n^{-1}) = O(n^{-1}\nu^\Theta),$$

wobei von (10) Gebrauch gemacht wurde.

Für die Fourierkoeffizienten gilt

$$\gamma_\nu = \frac{(-i)^\kappa}{2\pi\nu^\kappa} \int_{-\pi}^{\pi} f^{(\kappa)}(y)\, e^{-i\nu y}\, dy .$$

Setzt man das in (7) ein und vertauscht Integration und Summation, so folgt

$$R_n[f] = \frac{(-i)^\kappa}{2\pi} \int_{-\pi}^{\pi} f^{(\kappa)}(y) \sum_{\nu=-\infty}^{\infty} \rho_\nu\, \nu^{-\kappa} e^{-i\nu y}\, dy .$$

Der Teil des Integrals, der über das Intervall $I_1=[-\pi,\pi]-I_2$ erstreckt ist, wird nun mittels der Cauchy-Schwarzschen Ungleichung abgeschätzt durch

$$\{ \int_{I_1} [f^{(\kappa)}(y)]^2 dy \int_{I_1} | \sum_\nu \rho_\nu\, \nu^{-\kappa} e^{-i\nu y} |^2 dy\}^{1/2} \leq$$

$$O(n^{-1/2})\ \{ \int_{-\pi}^{\pi} |\ldots| dy\}^{1/2} = O(n^{-1/2})\ \{ \sum_\nu |\rho_\nu|^2\ \nu^{-2\kappa}\}^{1/2}$$

$$= O\ (\{n^{-1} \sum_{\nu=1}^{n-1} |b_\nu - c_{\nu,n}|^2\}^{1/2}) + O(n^{\Theta-\kappa}).$$

Der Integrand des verbleibenden Integrals wird umgeformt mittels der Identitäten

$$e^{-i\nu y} = -\frac{1}{4} \sin^{-2} \frac{y}{2}\ [e^{-i(\nu-1)y} - 2e^{-i\nu y} + e^{-i(\nu+1)y}]$$

und

$$\sum_{\nu=1}^{\infty} (e_{\nu-1} - 2e_\nu + e_{\nu+1}) r_\nu = e_0 r_1 - e_1 r_0 + \sum_{\nu=1}^{\infty} e_\nu (r_{\nu+1} - 2r_\nu + r_{\nu-1}).$$

Also

$$R[f] = - \frac{(-i)^\kappa}{8\pi} \int_{I_2} f^{(\kappa)}(y) \sin^{-2}\frac{y}{2} \sum_{|\nu|<n-1} e^{-i\nu y}\,[(\nu+1)^{-\kappa}\rho_{\nu+1} - 2\nu^{-\kappa}\rho_\nu$$

$$+(\nu-1)^{-\kappa}\rho_{\nu-1}]dy \; - \; \frac{(-i)^\kappa}{8\pi} \int_{I_2} \dots \sum_{|\nu|\geq n-1} \dots dy$$

$$+ O(\{n^{-1} \sum_{\nu=1}^{n-1} |b_\nu - c_{\nu,n}|^2\}^{1/2}) + O(n^{\Theta-\kappa})$$

$$(14) \qquad \equiv A_1 + A_2 + O(\dots) + O(\dots).$$

Für A_2 erhält man durch eine einfache Umformung des Integranden:

$$- \frac{(-i)^\kappa}{8\pi} \int_{I_2} f^{(\kappa)}(y) \sin^{-2}\frac{y}{2} \sum_{|\nu|\geq n-1} \nu^{-\kappa}(\rho_{\nu-1} - 2\rho_\nu + \rho_{\nu+1})dy$$

$$- \frac{(-i)^\kappa}{8\pi} \int_{I_2} f^{(\kappa)}(y) \sin^{-2}\frac{y}{2} \sum_{|\nu|\geq n-1} [\nu^{-\kappa} - (\nu+1)^{-\kappa}](\rho_{\nu-1} - \rho_{\nu+1})dy$$

$$- \frac{(-i)^\kappa}{8\pi} \int_{I_2} f^{(\kappa)}(y) \sin^{-2}\frac{y}{2} \sum_{|\nu|\geq n-1} [(\nu-1)^{-\kappa} - 2\nu^{-\kappa} + (\nu+1)^{-\kappa}]\rho_{\nu-1}dy.$$

Das dritte Integral wird abgeschätzt durch

$$\int_{I_2} |f^{(\kappa)}(y) \sin^{-2}\frac{y}{2}|dy \;\; O(\sum_{|\nu|\geq n-1} \nu^{-\kappa-2}|\rho_{\nu-1}|) =$$

$$O(n)\; O(\sum_{|\nu|\geq n-1} \nu^{-\kappa-2+\Theta}) = O(n^{\Theta-\kappa}).$$

Das zweite Integral wird ähnlich abgeschätzt: Für diejenigen Summanden, für die man (12) zur Verfügung hat, geht alles analog; für die restlichen verwendet man (11) und erhält ebenfalls $O(n^{\Theta-\kappa})$.

Im ersten Integral schätzt man wiederum nur die Summanden ab, für die (13) gilt, hier nach vorhergehender Anwendung der Cauchy-Schwarzschen Ungleichung.

Die verbleibende Summe ist

$$\sum_{\substack{\lambda=-\infty \\ \lambda\neq 0}}^{\infty} \sum_{\tau=-1}^{1} e^{-i(\lambda n+\tau)y}(\lambda n+\tau)^{-\kappa}[\rho_{\lambda n-1+\tau}-2\rho_{\lambda n+\tau}+\rho_{\lambda n+1+\tau}]$$

$$= \sum_{\substack{\lambda=-\infty \\ \lambda\neq 0}}^{\infty} e^{-i\lambda ny}(\lambda n)^{-\kappa}[e^{iy}(\rho_{\lambda n-2}-2\rho_{\lambda n-1}+\rho_{\lambda n}) + (\rho_{\lambda n-1}-2\rho_{\lambda n}$$

$$+ \rho_{\lambda n+1}) + e^{-iy}(\rho_{\lambda n}-2\rho_{\lambda n+1}+\rho_{\lambda n+2})] \qquad + O(n^{\Theta-\kappa-1})$$

$$= \sum_{\substack{\lambda=-\infty \\ \lambda\neq 0}}^{\infty} e^{-i\lambda ny}(\lambda n)^{-\kappa}[i \sin y(\rho_{\lambda n-2}-2\rho_{\lambda n-1}+2\rho_{\lambda n+1}-\rho_{\lambda n+2})$$

$$+ \cos y(\rho_{\lambda n-2}-\rho_{\lambda n-1}-\rho_{\lambda n+1}+\rho_{\lambda n+2}) +2\sin^2 \frac{y}{2}(\rho_{\lambda n-1}$$

$$- 2\rho_{\lambda n} + \rho_{\lambda n+1})] + O(n^{\Theta-\kappa-1}).$$

Den Term mit $\sin^2 \frac{y}{2}$ kann man im Integral leicht abschätzen (Cauchy-Schwarz und (11)) und erhält wieder $O(n^{\Theta-\kappa})$. Zur Diskussion der beiden anderen Summanden ist es nötig, zur Definition der ρ_ν zurückzukehren. Man erhält nach einiger Rechnung, wenn man den Faktor von i sin y mit s_λ und den von cos y mit c_λ bezeichnet:

$$s_\lambda = O(\lambda^\Theta n^{\Theta-2}) + \begin{cases} H_1 \quad -2i\ \mathrm{Im} c_1 & \text{wenn } \lambda \text{ gerade} \\ H_2 \quad +2i\ \mathrm{Im}(c_n e^{inx}) & \text{wenn } \lambda \text{ ungerade} \end{cases}$$

$$c_\lambda = O(\lambda^\Theta n^{\Theta-2}) + \begin{cases} H_3 & \text{wenn } \lambda \text{ gerade} \\ H_4 & \text{wenn } \lambda \text{ ungerade} \end{cases}$$

Dabei sind die H_τ von λ unabhängige Funktionen von x, für die $\|H_\tau\| = O(n^{\Theta-1})$ gilt.

Nun können die Standard-Abschätzungsmethoden wieder angewandt werden, man beachte dabei, daß

$$\sum_{\substack{\lambda=-\infty \\ \lambda \neq 0}}^{\infty} \frac{e^{-i\lambda ny}}{\lambda}$$

eine beschränkte Summe hat. Es bleibt schließlich übrig:

$$A_2 = \frac{(-i)^\kappa}{4\pi n^\kappa} \int_{I_2} f^{(\kappa)}(y) \sin^{-2} \frac{y}{2} \{-\sin y \sum_{\substack{\lambda=-\infty \\ \lambda \neq 0}}^{\infty} \frac{e^{-i2\lambda ny}}{(2\lambda)^\kappa} \mathrm{Im} c_1$$

$$+ \sin y\ \mathrm{Im}(c_n e^{inx}) \sum_{\lambda=-\infty}^{\infty} \frac{e^{-i(2\lambda+1)ny}}{(2\lambda+1)^\kappa} \} dy + O(n^{\Theta-\kappa})$$

$$= \frac{(-1)^\kappa \pi^{\kappa-1}}{4n^\kappa} \int_{I_2} f^{(\kappa)}(y) \operatorname{ctg} \frac{y}{2} \{2\ \mathrm{Im}(c_1) B_\kappa (\frac{ny}{\pi}) +$$

$$\mathrm{Im}(c_n e^{inx})\ E_{\kappa-1} (\frac{ny}{\pi})\} dy + O(n^{\Theta-\kappa}) .$$

Setzt man das in (14) ein, so folgt das Lemma.

3. Beweis der Sätze 1,2,4

Für alle hier betrachteten Spezialfälle (auch die Bernstein-Rogosinski-Modifikation) sind die Voraussetzungen des Lemmas erfüllt und es gilt sogar

$$R_n[f] = \frac{1}{4\pi} \left(\frac{-\pi}{n}\right)^{\kappa} \int_{I_2} f^{(\kappa)}(y) \operatorname{ctg}\frac{y}{2} \{2 \operatorname{Im}(c_1) B_{\kappa}\left(\frac{ny}{\pi}\right) +$$

(15)

$$\operatorname{Im}(c_n e^{inx}) E_{\kappa-1}\left(\frac{ny}{\pi}\right)\} \, dy \quad + \quad O(n^{\Theta-\kappa}).$$

Das ist nur im Fall des Satzes 2 nicht sofort zu sehen, hier ist zu zeigen

$$\left| \int_{I_2} f^{(\kappa)}(y) \sin^{-2}\frac{y}{2} \sum_{|\nu|<n-1} e^{-i\nu y} \left[(\nu+1)^{-\kappa}\rho_{\nu+1} - 2\nu^{-\kappa}\rho_{\nu} + (\nu-1)^{-\kappa}\rho_{\nu-1}\right] dy \right|$$

$$= O(n^{-\kappa})$$

mit

$$\rho_{\nu} = i \left(\frac{\nu}{n}\right)^{\sigma} e^{i\nu x} \operatorname{sgn} \nu.$$

Diese Beziehung folgt mit der schon oft benutzten Integralabschätzung aus

$$n^{-\sigma} \left| \sum_{\nu=1}^{n-2} e^{-i\nu(y-x)} \left[(\nu+1)^{\sigma-\kappa} e^{ix} - 2\nu^{\sigma-\kappa} + (\nu-1)^{\sigma-\kappa} e^{-ix}\right] \right| =$$

$$\Theta(n^{-\kappa-1})$$

und diese Gleichung ist richtig wegen

$$\left|(\nu+1)^{\sigma-\kappa} e^{ix} - 2\nu^{\sigma-\kappa} + (\nu-1)^{\sigma-\kappa} e^{-ix}\right| \leq \left[(\nu+1)^{\sigma-\kappa} - 2\nu^{\sigma-\kappa} + (\nu-1)^{\sigma-\kappa}\right]$$

$$+|\nu^{\sigma-\kappa}-(\nu+1)^{\sigma-\kappa}|\ |e^{-ix}-e^{ix}|\ +\nu^{\sigma-\kappa}|\ e^{-ix}-2+e^{ix}|$$

$$= O(n^{\sigma-\kappa-2}),$$

wobei noch von (10) Gebrauch gemacht ist. Der Integrand von (15) werde zur Abkürzung mit

$$f^{(\kappa)}(y)\ \mathrm{ctg}\ \frac{y}{2}\ h(\frac{ny}{\pi})$$

bezeichnet. Aus (15) folgt sofort

$$\sup_{\|f^{(\kappa)}\|\leq 1} |R_n[f]| \leq \frac{\pi^{\kappa-1}}{4n^\kappa} \int_{I_2} |\mathrm{ctg}\ \frac{y}{2}\ h(\frac{ny}{\pi})|\ dy + O(n^{\Theta-\kappa})$$

Hier gilt tatsächlich das Gleichheitszeichen, man wähle

$$f_{ext}^{(\kappa)}(x) = \begin{cases} 0 & x \in I_1, \\ \mathrm{sgn}\ \mathrm{ctg}\ \frac{y}{2}\ h(\frac{ny}{\pi}) & x \in I_2. \end{cases}$$

Dies ist auch wirklich Ableitung einer periodischen Funktion, denn ihr Mittelwert ist Null. Ist nämlich κ gerade, so ist $f_{ext}^{(\kappa)}$ ungerade und die Behauptung klar. Ist κ ungerade, so ist zu beachten, daß h die Periode $2\pi n^{-1}$ und den Mittelwert Null hat. Daraus folgt die Behauptung, wenn n ungerade ist, im anderen Fall ist die Definition von $f_{ext}^{(\kappa)}$ dahingehend zu modifizieren, daß sie in $[x_{-n}, x_{-n+1}] \cup [x_{n-1}, x_n]$ Null gesetzt wird. Das ergibt nur eine hier unwesentliche Änderung des Integralwertes um $O(n^{\Theta-\kappa})$.

Es ist nun zu berechnen

$$\frac{\pi^{\kappa-1}}{4n^\kappa} \int_{I_2} |\mathrm{ctg}\ \frac{y}{2}\ h(\frac{ny}{\pi})|dy = \frac{\pi^{\kappa-1}}{2n^\kappa} \int_{\pi n^{-1}}^{\pi} \mathrm{ctg}\ \frac{y}{2}|h(\frac{ny}{\pi})|dy$$

$$= \frac{\pi^\kappa}{2n^{\kappa+1}} \int_1^n \operatorname{ctg} \frac{\pi z}{2n} \, |h(z)| \, dz = \frac{\pi^\kappa}{2n^{\kappa+1}} \int_{-1}^{1} |h(z)| \sum_{\nu=1}^{[\frac{n-1}{2}]} \operatorname{ctg} \frac{2\nu\pi+\pi z}{2n} \, dz$$

$$+ O(n^{\Theta-\kappa}) = \frac{\pi^\kappa}{2n^{\kappa+1}} \sum_{\nu=1}^{[\frac{n-1}{2}]} \operatorname{ctg} \frac{2\nu\pi+\pi\xi}{2n} \int_{-1}^{1} |h(z)| \, dz + O(n^{\Theta-\kappa})$$

$$= \frac{\pi^{\kappa-1}}{2n^\kappa} \ln n \int_{-1}^{1} |h(z)| \, dz + O(n^{-\kappa+\Theta}) =$$

$$= \frac{\pi^{\kappa-1}}{n^\kappa} \ln n \int_0^1 |h(z)| \, dz + O(n^{-\kappa+\Theta}).$$

Bei dem letzten Schritt sind die Symmetrieverhältnisse von B_κ und $E_{\kappa-1}$ ausgenützt, man beachte insbesondere

$$\int_0^1 |\mathrm{const}_1 \, B_\kappa(z) + \mathrm{const}_2 \, E_{\kappa-1}(z)| dz = \int_0^1 |\mathrm{const}_1 B_\kappa(z)$$

$$- \mathrm{const}_2 E_{\kappa-1}(z)| dz.$$

Aus der erhaltenen Formel sind nun die Sätze 1,2,4 sofort abzulesen, abgesehen von der Ungleichung (4), die jetzt bewiesen werden soll.

Dazu wird gezeigt

$$(16) \qquad k(u) = \int_0^1 |uE_{\kappa-1}(z)+2B_\kappa(z)| dz$$

ist für $u\varepsilon[0,1]$ wachsend. κ möge gerade sein, im anderen Fall ist der Beweis etwas abzuändern. Eine Standard-Überlegung mit dem Satz von Rolle zeigt, daß der Integrand höchstens zwei Nullstellen haben kann. Wenn tatsächlich zwei Nullstellen $\xi_1=\xi_1(u)$ und $\xi_2=\xi_2(u)$ vorhanden sind, dann ist

$$(-1)^{\kappa/2}\ k(u) = -\int_0^{\xi_1} [\ uE_{\kappa-1}(z)+2B_\kappa(z)]dz + \int_{\xi_1}^{\xi_2}\ldots \int_{\xi_2}^{1}\ldots,$$

also

$$(-1)^{\kappa/2}\ k'(u) = -\int_0^{\xi_1} E_{\kappa-1}(z)dz + \int_{\xi_1}^{\xi_2} E_{\kappa-1}(z)dz - \int_{\xi_2}^{1} E_{\kappa-1}(z)dz =$$

$$= -\ 2\ E_\kappa(\xi_1) + 2E_\kappa(\xi_2).$$

Ist u=O, so ist $\frac{1}{2} - \xi_1 = \xi_2 - \frac{1}{2}$, läßt man u wachsen, so fallen (Skizze!) ξ_1 und ξ_2, also $\frac{1}{2} - \xi_1 > \xi_2 - \frac{1}{2}$ für $u > 0$. Demnach ist dann aus einer Skizze von E_κ abzulesen

$$\operatorname{sgn}[-E_\kappa(\xi_1)+E_\kappa(\xi_2)] = \operatorname{sgn} E_\kappa(\xi_2) = (-1)^{\kappa/2}\ ,$$

so daß schließlich $k'(u) > 0$ bewiesen ist.

Hat der Integrand von (16) nur eine Nullstelle, so kann man ähnlich argumentieren. Mindestens eine Nullstelle muß er haben, damit ist die Monotonie bewiesen.

Endlich ist in (4) noch ausgesagt

$$\int_0^1 |E_{\kappa-1}(z) + 2B_\kappa(z)|dz = 2^{\kappa+1}\int_0^1 |B_\kappa(z)|\ dz,$$

was man mit Hilfe der Fourierschen Reihen leicht nachrechnet.

4. Beweis von Satz 3

Ist f_o eine Funktion mit $\|f_o^{(\kappa)}\| \leq 1$, die an allen Stützstellen den Wert Null hat, so ist offensichtlich $|K[f_o](o) - \tilde{K}(o,\ldots,o)|$ eine untere Schranke für das gesuchte Supremum.

Mit Hilfe von $-f_o$ erhält man $|K[f_o](o) + \tilde{K}(o,\dots,o)|$ als weitere untere Schranke; aus beiden zusammen ergibt sich unser Ausgangspunkt

$$(17) \qquad \sup_{\|f^{(\kappa)}\| \leq 1} |K[f](o) - \tilde{K}[f(x_{-n+1},\dots,f(x_n)]| \geq |K[f_o](o)|$$

Nun ist bekannt, daß

$$K[f](o) = -\frac{1}{2\pi} \int_{-\pi}^{\pi} f(t)\operatorname{ctg}\frac{t}{2}\,dt$$

gilt, dabei ist das Integral als Cauchyscher Hauptwert zu verstehen. Wählt man nun für f eine ungerade Funktion, die auf $[x_{-1},x_1]$ identisch Null ist und auf $[x_1,\pi]$ periodisch mit der Periode πn^{-1}, so erhält man mit einer Rechnung, wie sie im vorigen Abschnitt schon verwendet wurde

$$|K[f](o)| = \frac{2n}{\pi^2}(\ln n + O(1))\left|\int_{x_1}^{x_2} f(t)dt\right|.$$

Setzt man weiter

$$f(t) = n^{-\kappa} g[n(t-\pi n^{-1})] \qquad t\varepsilon[x_1,x_2],$$

wobei g eine auf $]0,\pi[$ positive Funktion ist, die $\|g^{(\kappa)}\| \leq 1$ genügt und bei 0 und π je eine κ-fache Nullstelle hat, dann ist für dieses $f = f_o$

$$|K[f_o](o)| = \frac{2}{\pi^2}\,\frac{\ln n + O(1)}{n^\kappa}\int_0^\pi g(n)dn\;.$$

Mit (17) folgt nun sofort Satz 3.

5. Ein Beispiel

Als ein numerisches Beispiel sei $f(x)=|x|$ auf $[-\pi,\pi]$ gewählt. K[f] kann explizit angegeben werden:

$$K[f](x) = -\frac{2}{\pi} \int_0^x \ln|\operatorname{tg} \frac{y}{2}| \, dy .$$

Man erhält, jeweils auf 6 Dezimalen gerundet,

n =	10	100	1000
$R_n^{Wi}[f](x_1)$	-0,055167	-0,005408	-0,000541
$R_n^{Wi}[f](\frac{x_0+x_1}{2})$	-0,027174	-0,002704	-0,000270
$R_n^{BR}[f](x_1)$	0,007436	0,000730	0,000073
Satz 1	0,230259	0,046052	0,006908

Man erkennt, daß der Fehler in der Mitte zwischen den Stützstellen x_2 und x_1 nur etwa halb so groß ist wie bei x_1, was nach Satz 1 plausibel ist. Ferner sieht man, daß die Modifikation hier zu einem merklichen Genauigkeitsgewinn führt.

Die asymptotische Fehlerschranke aus Satz 1 ist nicht unrealistisch, führt aber zu einer mit n wachsenden Überschätzung des Fehlers. Der wahre Fehler geht hier nämlich wie n^{-1} gegen Null. Ein Blick auf (15) zeigt die Problematik bei diesen Abschätzungen: Diejenigen Funktionen, für die die hergeleitete Fehlerschranke angenommen wird, müssen eine stark oszillierende

κ - te Ableitung haben. Die "normalen" Funktionen sind gerade nicht von diesem Typ und daher wird die (unverbesserbare) Restschranke für sie zu einer Fehlerüberschätzung führen.

Literatur

[1] Gaier, D.: Konstruktive Methoden der konformen Abbildung; Springer-Verlag 1964

[2] Gutknecht, M.H.: Solving Theodorsen's integral equation for conformal maps with the fast Fourier transform; Seminar für Angewandte Mathematik ETH Zürich, Research Report 79 - 05 (1979)

[3] Haverkamp, R.: Approximationsgüte der Ableitungen bei trigonometrischer Approximation (Preprint Universität Münster 1981)

[4] Knauff, W. und Kreß , R.: Optimale Approximation linearer Funktionale auf periodischen Funktionen; Num. Math. 22, S. 187 - 205 (1974)

[5] Meinardus, G.: Approximation von Funktionen und ihre numerische Behandlung; Springer Verlag 1964

[6] Nikolsky, S.M.: An asymptotic estimation of the remainder under approximation by interpolating trigonometric polynomials; C.R. (Doklady) de l'Académie des Sciences de l'URSS 31, S. 215 - 218 (1941)

[7] Onegov, L.A.: Quadraturformeln für singuläre Integrale (Russ.); Izv. Vyss. Ucebn Zaved. Matematika 1978, no. 4 (191), S. 64 - 78 (1978)

[8] Zygmund, A. : The approximation of functions by typical means of their Fourier series; Duke Math. J. 12, S. 695 - 704 (1945)

Prof. Dr. Helmut Braß, Lehrstuhl E für Mathematik
der Technischen Universität Braunschweig,
3300 Braunschweig, Pockelsstraße 14, BRD

ON A SPECIAL LAURENT-HERMITE INTERPOLATION PROBLEM

Adhemar Bultheel

We present a recursive algorithm for the construction of a rational approximation for a given Laurent series which in a certain sense interpolates at the zeros of its numerator. If certain symmetry conditions are satisfied, the algorithms of Nevanlinna-Pick and Schur are found as special cases. We give also an interpolation of the algorithm as a coupled recursion for reproducing kernels of indefinite inner product spaces defined with the aid of the given Laurent series. In the symmetric case, the approximation can be given a least squares interpretation. The interpolation points can then be chosen in an optimal way. A numerical example of the latter problem is given.

1. Problem formulation

The Laurent-Hermite interpolation problem as we define it here is an attempt to generalize the concept of Laurent-Padé approximation to a related interpolation problem.

We introduce the definition gradually and start with the non-Laurent problem :

Suppose F(x) is an analytic function in a region of the complex plane (suppose it is the unit disc) and let $\alpha_0, \alpha_1, \alpha_2, \ldots$ be a sequence of points (some of them may coincide) in this region. The (rational) Hermite interpolation problem consists in finding two polynomials $P^{m,n}$ and $Q^{m,n}$ of degrees bounded by m and n respectively such that $R^{m,n} = P^{m,n}/Q^{m,n}$ interpolates F in $\alpha_0, \alpha_1, \ldots \alpha_N$ with $N \geq m+n$.

Thus

$$[F - R^{m,n}](x) = g(x) \prod_0^N (x-\alpha_i)$$

with g analytic in the disc. If all points α_i are different, this is the Cauchy problem. If all points $\alpha_i = 0$, this is the Padé problem.

For the Laurent-Padé problem F(x) need not be analytic everywhere inside the unit disc. Suppose it has a Laurent series expansion around x = 0 valid in a region containing the unit circle $|x| = 1$. Then one tries to find <u>trigonometric</u> polynomials $P^{m,n}$ and $Q^{m,n}$ of degree not exceeding m and n respectively such that $R^{m,n} = P^{m,n}/Q^{m,n}$ has a Laurent series expansion in the neighborhood of $|x| = 1$ being such that $F - R^{m,n}$ contains no terms in x^i for $|i| \leq N$.

This problem has received some attention [11,6]. It is known to be related to the Schur algorithm [1,16] and thus to two-point Padé approximation [17]. This means that one constructs a rational function approximating two power series simultaniously, viz. the "analytic part" of F and the "coanalytic part" of F. One part is approximated at the origin, the other at infinity.

The Laurent-Hermite interpolation problem is a direct generalization of this. We interpolate the "analytic part" at $0 = \alpha_0, \alpha_1, \alpha_2, \ldots$, a sequence of points inside the unit disc and the "coanalytic part" at points $\infty = 1/\bar{\beta}_0, 1/\bar{\beta}_1, 1/\bar{\beta}_2, \ldots$ outside the unit circle (upper bar denotes complex conjugation) so that we have in the end

$$F(x) - R^{m,n}(x) = \Delta_1^{m,n}(x) \prod_0^N (x-\alpha_i) + \Delta_2^{m,n}(1/x) \prod_0^N \left(\frac{1}{x} - \bar{\beta}_i\right) \tag{1}$$

with $\Delta_1^{m,n}(x)$ and $\Delta_2^{m,n}(1/x)$ as functions of x, analytic inside the unit disc, everywhere, except for poles of $R^{m,n}$.

In this paper we are not going to consider this general case but an interesting special version of it. Suppose we know the zeros of F, or at least, can estimate them. If they are $\alpha_1, \alpha_2, \ldots$ and $1/\bar{\beta}_1, 1/\bar{\beta}_2, \ldots$ with $|\alpha_i|, |\beta_i| < 1$, then it is wise to propose an approximant $R^{m,n}$ of the form

$$R^{m,n}(x) = \frac{\prod_1^m (x-\alpha_i)}{\psi_n(x)} \cdot \frac{\prod_1^m (1/x-\bar{\beta}_i)}{\phi_{n\star}(1/x)} \tag{2}$$

with $\phi_{n\star}$ and ψ_n polynomials of degree at most n. If the estimates of α_i and β_i are exact and if F(x) is of the form (2), then we shall formulate and algorithm that gives $R^{m,n}(x) = F(x)$ after max(m,n) steps. If the estimates α_i and β_i are not exact, then our procedure still provides an approximation in a restricted Laurent-Hermite sense. Of course we can not go as far as in (1) because we fix the numerator beforehand, thus reducing the degrees of freedom. However all remaining parameters in the denominator are used to satisfy (1) as far as possible. More precisely we require ψ_n and $\phi_{n\star}$ to be such that

$$F(x) - \frac{\prod_1^m (x-\alpha_i)(1/x-\bar{\beta}_i)}{\psi_n(x)\phi_{n\star}(1/x)} = \Delta_1^{m,n}(x)\prod_0^m (x-\alpha_i)/\psi_n(x) + \Delta_2^{m,n}(1/x)\prod_0^m (1/x-\bar{\beta}_i)/\phi_{n\star}(1/x)$$

with as before $\alpha_0 = \beta_0 = 0$ and $\Delta_1^{m,n}(x)$ and $\Delta_2^{m,n}(1/x)$ as functions of x are analytic inside the unit disc.

First remark that if $m < n$, then we can always add $\alpha_{m+1} = \alpha_{m+2} = \ldots = \alpha_n = \beta_{m+1} = \beta_{m+2} = \ldots = \beta_n = 0$ because this does not change the numerator. The case $m > n$ gives more trouble. We don't like to consider it at this moment and suppose that in this case m-n zeros are simply devided out of F(x) so that we restrict ourselves from now on to the case $m = n$.

2. Justification

The Laurent-Hermite interpolation problem has recently found a very interesting application in the construction of least squares ARMA (auto-regressive, moving average) filters [2, 3, 4, 7, 8, 9]. In this special case the given Laurent series $F(x) = \sum_{-\infty}^{\infty} f_k x^k$ represents an autocorrelation function and therefore $f_{-k} = \bar{f}_k$ and for all n the Toeplitz matrices constructed on the parameters $\{f_j\}_{j=-n}^{n}$ are positive definite. For the general theory we refer to the literature where it is shown how the classical Wiener-Massani AR-prediction theory [18] can be nicely adapted in this context. To explain the basic idea however we may restrict ourselves to the case that F(x) is rational. F(x) is then factorizable as $F(x) = s(x).\overline{s(1/\bar{x})}$ with s(x) rational having all its poles and zeros inside the unit disc. s(x) is in essence the filter one wants to construct from the covariance data F(x) and the location of the poles and zeros of s(x) are crucial for stability reasons. The example shows also why we do not consider the general Laurent-Hermite problem. Indeed the result of this would deliver an approximant to F(x) that may not be factorizable in general (i.e. not be positive on the unit circle) and thus gives no approximation to s(x) which is the ultimate goal. By restricting ourselves to the special problem described before, we fix the numerator as a product of linear factors with zeros separated as desired and the denominator is found as a product of two polynomials. Thus we directly find the approximant of s(x) because the zeros of the denominator are automatically separated. This results from the properties of F(x) and the algorithm we are going to define in the next section. Just as in the classical case of AR filtering, the approximant can be shown to be optimal in a weighted least squares sense. At the end of the paper we shall give a numerical example illustrating these ideas. The remaining sections give a general framework for these techniques. We give in section 3 the description of an algorithm in the style of Schur [1,16] and Nevanlinna-Pick [1] to construct the approximant recursively. Section 4 defines reproducing kernels for a certain indefinite inner product space and derives a Christoffel-Darboux type formula. In section 5 it is briefly sketched how these reproducing kernels satisfy a recursion that turns out to be exactly the same as the one described in section 3.

3. A Schur-like algorithm

In this section we describe an algorithm that shall give the required approximant recursively. It is a continued fraction type of algorithm which is a generalization of a similar recursive scheme proposed by Shur [1,16].

Given $F(x) = \sum_{-\infty}^{\infty} f_k x^k$, $f_0 = 1$, we split this up into two parts :

$F^+(x) = 1 + 2\sum_1^{\infty} f_k x^k$ and $F^-(1/x) = 1 + 2\sum_1^{\infty} f_k x^k$, thus $F(x) = \frac{1}{2}[F^-(1/x)+F^+(x)]$

We call $\frac{1}{2}[1+F^+(x)]$ the analytic part of $F(x)$ and $-\frac{1}{2}[1-F^-(1/x)]$ the coanalytic part of $F(x)$.

We construct the matrix

$$\Delta_0(x) = \frac{1}{2}\begin{bmatrix} 1+F^+(x) & -(1-F^+(x)) \\ -(1-F^-(1/x)) & 1+F^-(1/x) \end{bmatrix} \overset{\text{def}}{=} \begin{bmatrix} \Delta_{11}^0(x) & \Delta_{12}^0(x) \\ \Delta_{21}^0(1/x) & \Delta_{22}^0(1/x) \end{bmatrix}$$

Now we successively transform $\Delta_0(x)$ into $\Delta_1(x), \Delta_2(x), \ldots,$ such that

$$\Delta_n(x) = \begin{bmatrix} \Delta_{11}^n(x) & \Delta_{12}^n(x) \\ \Delta_{21}^n(x) & \Delta_{22}^n(1/x) \end{bmatrix}$$

satisfies :

$$\begin{aligned} \Delta_{11}^n(x) &= \prod_1^n (x-\alpha_i) . \tilde{\Delta}_{11}^n(x) \quad ; \quad \Delta_{12}^n(x) = \prod_0^n (x-\alpha_i)\tilde{\Delta}_{12}^n(x) \\ \Delta_{21}^n(1/x) &= \frac{1}{x}\prod_1^n (1-\bar{\beta}_i x) . \tilde{\Delta}_{21}^n(1/x) \quad ; \quad \Delta_{22}^n(1/x) = \prod_1^n (1-\bar{\beta}_i x)\tilde{\Delta}_{22}^n(1/x) \end{aligned} \tag{3}$$

with $\tilde{\Delta}_{1j}^n(x)$ and $\tilde{\Delta}_{2j}^n(1/x)$ as functions of x analytic inside the unit disc.

Clearly these requirements are met for n = 0.

The transform from $\Delta_{n-1}(x)$ to $\Delta_n(x)$ is described as follows :

$$\Delta_n(x) = \Delta_{n-1}(x)\theta_n(x)$$

with

$$\theta_n(x) = \begin{bmatrix} 1 & -\gamma_n^\alpha \\ -\overline{\gamma_n^\beta} & 1 \end{bmatrix} \begin{bmatrix} x-\alpha_n & 0 \\ 0 & 1-\bar{\beta}_n x \end{bmatrix} \begin{bmatrix} 1 & -\rho_n^\alpha \\ -\overline{\rho_n^\beta} & 1 \end{bmatrix}$$

$$\rho_n^\alpha = \tilde{\Delta}_{12}^{n-1}(\alpha_n)/\tilde{\Delta}_{11}^{n-1}(\alpha_n) \quad ; \quad \gamma_n^\alpha = \alpha_n\,\rho_n^\alpha$$

$$\overline{\rho_n^\beta} = \tilde{\Delta}_{21}^{n-1}(\bar{\beta}_n)/\tilde{\Delta}_{22}^{n-1}(\bar{\beta}_n) \quad ; \quad \gamma_n^\beta = \beta_n\,\rho_n^\beta$$

if α_n and $\beta_n \neq 0$. For $\alpha_n = 0$ or $\beta_n = 0$ we would have $\rho_n^\alpha = \rho_n^\beta = 0$. In that case we have to replace the formulas for ρ_n^α and ρ_n^β by

$$\rho_n^\alpha = \frac{d}{dx}\Gamma_n^\alpha(x)\Big|_{x=0}, \quad \gamma_n^\alpha = 0, \quad \Gamma_n^\alpha(x) = \tilde{\Delta}_{12}^{n-1}(x)/\tilde{\Delta}_{11}^{n-1}(x)$$

and

$$\overline{\rho_n^\beta} = \frac{d}{dx}\Gamma_n^\beta(x)\Big|_{x=0}, \quad \gamma_n^\beta = 0, \quad \Gamma_n^\beta(x) = \tilde{\Delta}_{21}^{n-1}(x)/\tilde{\Delta}_{22}^{n-1}(x)$$

It can be verified that $\rho_n^{\alpha/\beta}$ and $\gamma_n^{\alpha/\beta}$ are chosen such that if $\Delta_{n-1}(x)$ satisfies (3) with n replaced by (n-1), then also $\Delta_n(x)$ will satisfy (3).

Now consider the matrix

$$\Theta_n(x) = \theta_1(x)\theta_2(x)\ \ldots\ \theta_n(x).$$

It is clear from its construction that the elements of $\Theta_n(x)$ are polynomials in x of degree n. By taking linear combinations of elements in $\Theta_n(x)$ we define the polynomials $f_n^\star, g_n, \psi_n$ and $\phi_n^\star$ by the relation

$$\begin{bmatrix} 1 & 1 \\ 1 & -1 \end{bmatrix} \Theta_n(x) = \begin{bmatrix} \phi_n^\star & \psi_n \\ f_n^\star & -g_n \end{bmatrix}$$

or equivalently

$$\Theta_n(x) = \frac{1}{2}\begin{bmatrix} \phi_n^{\star} + f_n^{\star} & \psi_n - g_n \\ \phi_n^{\star} - f_n^{\star} & \psi_n + g_n \end{bmatrix} .$$

Multiplying out the product $\Delta_0(x)\Theta_n(x) = \Delta_n(x)$ where $\Delta_n(x)$ has the properties (3) gives for the (1,2) element :

$$-g_n(x) + F^+(x)\psi_n(x) = x \prod_1^n (x-\alpha_i)\tilde{\Delta}_{12}(x)$$

and for the (2,1) element :

$$-f_n^{\star}(x) + F^-(1/x)\phi_n^{\star}(x) = \frac{1}{x}\prod_1^n (1-\bar{\beta}_i x)\tilde{\Delta}_{21}(1/x)$$

or if we divide by x^n and set $f_n^{\star}(x)/x^n = f_{n\star}(1/x)$ and $\psi_n^{\star}(x)/x^n = \psi_{n\star}(1/x)$:

$$-f_{n\star}(1/x) + F^-(1/x)\phi_{n\star}(1/x) = \frac{1}{x}\prod_1^n (\frac{1}{x} - \bar{\beta}_i)\tilde{\Delta}_{21}(1/x).$$

Now

$$\det \Theta_n(x) = \prod_1^n \det \theta_i(x)$$

which gives

$$\phi_n^{\star}(x)g_n(x) + f_n^{\star}(x)\psi_n(x) = \prod_1^n (x-\alpha_i)(1-\bar{\beta}_i x)(1-\gamma_i^{\alpha}\overline{\gamma_i^{\beta}})(1-\rho_i^{\alpha}\overline{\rho_i^{\beta}})$$

$$= 2C_n \prod_1^n (x-\alpha_i)(1-\bar{\beta}_i x)$$

where C_n is a constant.

If $\psi_n(x)$ and $\phi_{n\star}(1/x)$ have no zeros on the unit circle, then in a certain region containing this circle we may define

$$F_n^+(x) = g_n(x)/\psi_n(x) \quad \text{and} \quad F_n^-(1/x) = f_{n\star}(1/x)/\phi_{n\star}(1/x)$$

and

$$F_n(x) = \frac{1}{2}[F_n^-(1/x) + F_n^+(x)] = C_n \frac{\prod_1^n (x-\alpha_i)(\frac{1}{x} - \bar{\beta}_i)}{\psi_n(x)\phi_{n\star}(1/x)} .$$

We have

$$F(x) - F_n(x) = \prod_0^n (x-\alpha_i)\tilde{\Delta}_{12}(x)/\psi_n(x) + \prod_0^n (\frac{1}{x} - \bar{\beta}_i)\tilde{\Delta}_{21}(\frac{1}{x})/\phi_{n\star}(\frac{1}{x})$$

as required.

This completes the description of the algorithm.

4. Indefinite inner product spaces, reproducing kernels and Christoffel-Darboux relations

Like the Schur algorithm is related to the Szegö polynomials orthogonal on the unit circle, the algorithm of the previous is related to orthogonal (rational) functions and reproducing kernels for some indefinite inner product space which we shall introduce in this section.

Consider the space $\mathcal{F}$ of functions having a Laurent series expansion in a region containing the unit circle. We define two subspaces of rational functions of a certain degree n with given poles outside the unit disc :

$$\mathcal{L}_n^{\alpha} = \{ \frac{p_n(x)}{\prod_1^n (1-\bar{\alpha}_j x)} \text{ , } p_n \text{ polynomial of degree } n, |\alpha_j| < 1 \}$$

$$\mathcal{L}_n^{\beta} = \{ \frac{q_n(x)}{\prod_1^n (1-\bar{\beta}_j x)} \text{ , } q_n \text{ polynomial of degree } n, |\beta_j| < 1 \}$$

For fixed $F(x) = \sum_{-\infty}^{\infty} f_k x^k$, $f_0 = 1$, we define a linear functional C such that

$$C(x^k) = f_{-k} \qquad k = \ldots,-1,0,1,2,\ldots$$

and we associate with it the following indefinite inner product over $\mathcal{F}$

$$\langle f,g \rangle = C(f\, g_\star)$$

where by definition $h_\star(x) = \overline{h(1/\bar{x})}$.

We define also transformations from $\mathcal{L}_n^{\alpha}$ into $\mathcal{L}_n^{\beta}$ and conversely by

$$f_n \in \mathcal{L}_n^{\alpha} \to f_n^{\alpha\beta} \overset{\text{def}}{=} U_n^{\alpha\beta}(x) f_{n\star}(x) \in \mathcal{L}_n^{\beta}$$

$$g_n \in \mathcal{L}_n^{\beta} \to g_n^{\beta\alpha} \overset{\text{def}}{=} U_n^{\beta\alpha}(x) g_{n\star}(x) \in \mathcal{L}_n^{\alpha}$$

where

$$U_n^{\alpha\beta}(x) = \prod_{i=1}^{n} \frac{x-\alpha_i}{1-\bar{\beta}_i x} \quad \text{and} \quad U_n^{\beta\alpha}(x) = \prod_{i=1}^{n} \frac{x-\beta_i}{1-\bar{\alpha}_i x}$$

With these transformations we have the following relations as can be easily verified

PROPERTY 4.1 For $f \in \mathcal{L}_n^{\alpha}$ and $g \in \mathcal{L}_n^{\beta}$:

$$<f,g> = <g_\star, f_\star> = <g^{\beta\alpha}, f^{\alpha\beta}>$$

The recursive algorithm of the previous section is closely related to the construction of reproducing kernels [14] for $\mathcal{L}_n^{\alpha}$ and $\mathcal{L}_n^{\beta}$, $n = 0,1,\ldots$. These kernels are defined in the following way :

The kernels $k_n(x,y)$ and $\ell_n(x,y)$ are in $\mathcal{L}_n^{\beta}$ resp. $\mathcal{L}_n^{\alpha}$ as a function of x for fixed y and are such that

$$<f(.),k_n(.,y)> = f(y) \qquad \forall f \in \mathcal{L}_n^{\alpha}$$

$$<\ell(.,y),g(.)> = \overline{g(y)} \qquad \forall g \in \mathcal{L}_n^{\beta}.$$

As in the classical case, under certain non-degeneracy conditions for the product space, (which we suppose to be satisfied wherever needed) these kernels are uniquely defined and can be expressed as a combination of a biorthogonal basis for $\mathcal{L}_n^{\alpha}$ and $\mathcal{L}_n^{\beta}$.

Indeed, suppose $\{\lambda_j\}_0^n$ and $\{\mu_j\}_0^n$ form a basis for $\mathcal{L}_n^{\alpha}$ and $\mathcal{L}_n^{\beta}$ respectively and are biorthonormal in the sense that $<\lambda_i,\mu_j> = \delta_{ij}$. It is easily verified that

$$k_n(x,y) = \sum_0^n \overline{\lambda_j(y)}\ \mu_j(x) \quad \text{and} \quad \ell_n(y,x) = \sum_0^n \overline{\mu_j(x)}\ \lambda_j(y)$$

LEMMA 4.2

$$\ell_n(y,x) = \overline{U_n^{\beta\alpha}(x)}\ U_n^{\beta\alpha}(y)\ \overline{k_n(1/\bar{y},1/\bar{x})}$$

PROOF

By definition is $<f(x),k_n(x,y)> = f(y)$, $\forall f \in \mathcal{L}_n^{\alpha}$. With property 4.1 then also

$$<U_n^{\beta\alpha}(x)\ \overline{k_n(1/\bar{x},y)},\ U^{\alpha\beta}(x)\ \overline{f(1/\bar{x})}> = f(y).$$

Take the ${}^{\alpha\beta}$ transform with respect to y then

$$<\overline{U_n^{\alpha\beta}(y)}\ U_n^{\beta\alpha}(x)\ \overline{k_n(1/\bar{x},1/\bar{y})},\ f^{\alpha\beta}(x)> = \overline{f^{\alpha\beta}(y)}.$$

If $f \in \mathcal{L}_n^{\alpha}$ is arbitrary then $f^{\alpha\beta} \in \mathcal{L}_n^{\beta}$ is arbitrary so that the result follows by definition of $\ell_n(x,y)$.

Using the biorthonormal basis functions, the following corollary is simple to prove

COROLLARY 4.3

$$\ell_n(y,\alpha_n) = \kappa_n^\lambda \mu_n^{\beta\alpha}(y) \;,\quad \ell_n(\beta_n,x) = \overline{\kappa_n^\mu}\;\overline{\kappa_n^{\alpha\beta}(x)}$$

and

$$\ell_n(\beta_n,\alpha_n) = \kappa_n^\lambda \overline{\kappa_n^\mu}$$

with

$$\kappa_n^\lambda = \overline{\lambda_n^{\alpha\beta}(\alpha_n)} \quad \text{and} \quad \kappa_n^\mu = \overline{\mu_n^{\beta\alpha}(\beta_n)}$$

The following theorem gives the analogue of the Christoffel-Darboux relation for orthogonal polynomials, now generalized to a relation for the biorthonormal systems.

THEOREM 4.4

$$k_n(x,y) = \frac{\overline{\mu_{n+1}^{\beta\alpha}(y)}\,\lambda_{n+1}^{\alpha\beta}(x) - \overline{\lambda_{n+1}(y)}\,\mu_{n+1}(x)}{1-\left(\overline{\dfrac{y-\beta_{n+1}}{1-\bar{\alpha}_{n+1}y}}\right)\left(\dfrac{x-\alpha_{n+1}}{1-\bar{\beta}_{n+1}x}\right)}$$

PROOF

Call the right hand side $R(x,y)$, then we have to prove that $\langle f(x),R(x,y)\rangle = f(y)$, $\forall f \in \mathcal{L}_n^\alpha$. Now $\langle f(x),R(x,y)\rangle = f(y)\langle 1,R(x,y)\rangle + \langle (x-y)h(x),R(x,y)\rangle$ where $f(x) - f(y) \overset{\text{def}}{=} (x-y)h(x) \in \mathcal{L}_n^\alpha$.

We first show that the second term vanishes. Indeed, working on the denominator of $R(x,y)$ we find with some algebra that

$$\langle (x-y)h(y),R(x,y)\rangle = c\,\langle h_1(x)(1-\bar{\alpha}y),\overline{\mu_{n+1}^{\beta\alpha}(y)}\,\lambda_{n+1}^{\alpha\beta}(x) - \overline{\lambda_{n+1}(y)}\,\mu_{n+1}(x)\rangle$$

with $h_1(x) = (x-\beta_{n+1})h(x) \in \mathcal{L}_n^\alpha$. Using property 4.1 and the orthogonality of μ_{n+1} and λ_{n+1} on $\mathcal{L}_n^\alpha$ and $\mathcal{L}_n^\beta$ respectively, it follows that this is zero.

We thus may conclude that $\langle f(x),R(x,y)\rangle = f(y)\eta(y)$ with $\eta(y) = \langle 1,R(.,y)\rangle$. Similarly, we can show that $\langle R(x,y),f(y)\rangle = \eta'(x)\overline{f(x)}$ where $\eta'(x) = \overline{\langle R(x,.)},1\rangle$. Using both results on $\langle R(x,.),R(.,y)\rangle$ we find that it equals $\eta(y)\overline{R(x,y)}$ and also $\eta'(x)\overline{R(x,y)}$ so that we must have $\eta'(x) = \eta(y) = \text{constant} = \eta$.

Take $x = \alpha_{n+1}$ and $y = \beta_{n+1}$, then we have shown that

$$\eta\, k_n(\alpha_{n+1},\beta_{n+1}) = \kappa_{n+1}^\mu \overline{\kappa_{n+1}^\lambda} - \overline{\lambda_{n+1}(\beta_{n+1})}\,\mu_{n+1}(\alpha_{n+1})$$

The first term in the right hand side equals $k_{n+1}(\alpha_{n+1},\beta_{n+1})$ because of corollary 4.3. So the difference in the right hand side is $k_n(\alpha_{n+1},\beta_{n+1})$ and thus is $\eta = 1$.

5. Recurrence relations

In this section we give a recursion for the reproducing kernels defined in the previous section and briefly indicate how it is related to the recursion in section 3. The proof of the correspondence between both recursions requires more knowledge about the indefinit inner product space which we shall not give in detail. More on this will be published later (see also [2,7,9]).

The reproducing kernels satisfy the recursions given in the following

THEOREM 5.1

$$(1-\rho^{\alpha}_{n+1}(y)\overline{\rho^{\beta}_{n+1}(y)})[\ell^{\alpha\beta}_{n+1}(x,y)k_{n+1}(x,y)] = [\ell^{\alpha\beta}_{n}(x,y)k_{n}(x,y)].$$

$$\begin{bmatrix} 1 & \left(\frac{y-\alpha_{n+1}}{1-\bar{\beta}_{n+1}y}\right)\rho^{\alpha}_{n+1}(y) \\ \overline{\left(\frac{y-\beta_{n+1}}{1-\bar{\alpha}_{n+1}y}\right)}\,\overline{\rho^{\beta}_{n+1}(y)} & 1 \end{bmatrix} \begin{bmatrix} \frac{x-\alpha_{n+1}}{1-\bar{\beta}_{n+1}x} & 0 \\ 0 & 1 \end{bmatrix} \begin{bmatrix} 1 & -\rho^{\alpha}_{n+1}(y) \\ -\overline{\rho^{\beta}_{n+1}(y)} & 1 \end{bmatrix}$$

with $\rho^{\alpha}_{n+1}(y) = -\mu_{n+1}(y)/\overline{\mu^{\beta\alpha}_{n+1}(y)}$

$$\rho^{\beta}_{n+1}(y) = -\lambda_{n+1}(y)/\overline{\lambda^{\alpha\beta}_{n+1}(y)}$$

PROOF

Obviously

$$k_{n+1}(x,y) - k_n(x,y) = \overline{\lambda_{n+1}(y)}\ \mu_{n+1}(x). \tag{4}$$

$\overline{\lambda_{n+1}(y)}\ \mu_{n+1}(x)$ is herein replaced by an expression that can be found from the Christoffel-Darboux formula. We obtain

$$k_{n+1}(x,y) = \overline{\left(\frac{y-\beta_{n+1}}{1-\bar{\alpha}_{n+1}y}\right)}\left(\frac{x-\alpha_{n+1}}{1-\bar{\beta}_{n+1}x}\right)k_n(x,y) + \overline{\mu^{\beta\alpha}_{n+1}(y)}\ \lambda^{\alpha\beta}_{n+1}(x) . \tag{5}$$

The $^{\alpha\beta}$ transform of a relation like (4) written down for $\ell_{n+1}(x,y)$ gives

$$\ell^{\alpha\beta}_{n+1}(x,y) = \left(\frac{x-\alpha_{n+1}}{1-\bar{\beta}_{n+1}x}\right)\ell^{\alpha\beta}_n(x,y) + \lambda^{\alpha\beta}_{n+1}(x)\ \mu_{n+1}(y) . \tag{6}$$

Extract $\lambda^{\alpha\beta}_{n+1}(x)$ from (6) and substitute in (5), then

$$\rho^{\alpha}_{n+1}(y)k_{n+1}(x,y) = \overline{\left(\frac{y-\beta_{n+1}}{1-\bar{\alpha}_{n+1}y}\right)}\left(\frac{x-\alpha_{n+1}}{1-\bar{\beta}_{n+1}x}\right)k_n(x,y)\ \rho^{\alpha}_{n+1}(y)$$

$$- \left[\ell^{\alpha\beta}_{n+1}(x,y) - \left(\frac{x-\alpha_{n+1}}{1-\bar{\beta}_{n+1}x}\right)\ell^{\alpha\beta}_n(x,y)\right] \tag{7}$$

with $\rho^{\alpha}_{n+1}(y) = -\mu_{n+1}(y)/\overline{\mu^{\beta\alpha}_{n+1}(y)}$.

Repeat the same thing for $\ell_{n+1}(x,y)$ and take the $^{\alpha\beta}$ transform. This yields :

$$\overline{\rho^{\beta}_{n+1}(y)}\,\ell^{\alpha\beta}_{n+1}(x,y) = \overline{\rho^{\beta}_{n+1}(y)}\left(\frac{y-\alpha_{n+1}}{1-\bar{\beta}_{n+1}y}\right)\ell^{\alpha\beta}_n(x,y) - k_{n+1}(x,y) + k_n(x,y) \tag{8}$$

with $\rho^{\beta}_{n+1}(y) = -\lambda_{n+1}(y)/\overline{\lambda^{\alpha\beta}_{n+1}(y)}$.

Combining (7) and (8) gives the recursion.

It follows simply from Theorem 5.1 that

COROLLARY 5.2

$$(1-\rho^{\alpha}_{n+1}(0)\overline{\rho^{\beta}_{n+1}(0)})k_{n+1}(0,0) = (1-\alpha_{n+1}\bar{\beta}_{n+1}\rho^{\alpha}_{n+1}(0)\overline{\rho^{\beta}_{n+1}(0)})k_n(0,0)$$

and because $k_0(0,0) = 1$:

$$k_n(0,0) = \prod_{1}^{n} (1-\alpha_i\bar{\beta}_i\rho^{\alpha}_i(0)\overline{\rho^{\beta}_i(0)})/(1-\rho^{\alpha}_i(0)\overline{\rho^{\beta}_i(0)}) .$$

There is a certain similarity between the $\theta_n(x)$ matrices of section 3 and the recursion for the reproducing kernels given above. To show the correspondence more explicitly, take $y = 0$ and suppose that $\phi^{*}_n(x)$ and $\psi_n(x)$ are the numerators of $\ell^{\alpha\beta}_n(x,0)c_n$ and $k_n(x,0)c_n$ respectively, with

$$c_n = \prod_{1}^{n} (1-\rho_i^{\alpha}(0)\overline{\rho_i^{\beta}(0)})$$

The recursion for these polynomials is then

$$[\phi_n^{\star}(x) \;\; \psi_n(x)] = [\phi_{n-1}^{\star}(x) \;\; \psi_{n-1}(x)]\theta_n(x) \tag{9}$$

with $\theta_n(x)$ of exactly the same form as the one defined in section 3. Only the definition of ρ_n^{α} and ρ_n^{β} is different.
The explicit identification of both recursions requires a more detailed study of the indefinite inner product space. If this space is not degenerate [5] a whole projection theory can be set up and the proofs are relatively simple. Consult [2,7,9] for similar proofs. If $F(x)$ is such that the inner product space is degenerate, then the algorithms may break down and we have a situation similar to the construction of Padé approximants for a function with non-normal Padé table. However the algorithms may be modified to overcome this situation and proofs go through, although everything becomes considerably more complicated.

That we found in section 3 only the recursion of the reproducing kernels for the special case $y = 0$ is due to the fact that we normalized in the Schur algorithm in each step at $x = 0$, and not in a general point $x = y$. If the algorithm of section 3 would have been adopted as such, then we would have found the complete recursion of theorem 5.1.

Thus both algorithms are completely equivalent. The algorithm of section 3 is called of <u>outgoing</u> type because it starts with the whole information about $F(x)$ in $\Delta_0(x)$ and at every step, some information $(\rho_i^{\alpha},\rho_i^{\beta})$ is extracted and the algorithm goes on with what "remains" of $F(x)$. This algorithm resembles much the division algorithms constructing continued fractions whose convergents are Padé approximants. The algorithm of this section is called of <u>incoming</u> type because one starts with no information $(\phi_0^{\star} = \psi_0 = 1)$ at all and gradually, as the algorithm comes along, more and more information about $F(x)$ is thrown in. This is accumulated in the polynomials of increasing degree. The matrix $\Theta_n(x)$ defined in section 3 contains also the polynomials $f_n^{\star}$ and g_n. These are related to reproducing kernels as $\phi_n^{\star}$ and ψ_n are. It are the kernels for similar rational function spaces, now with an indefinit inner product related to the function $\Phi(x) = \frac{1}{2}[\Phi^{-}(1/x) + \Phi^{+}(x)]$ with $\Phi^{+}(x)$ and $\Phi^{-}(x)$ the formal inverses of

$F^+(x)$ and $F^-(x)$ respectively.

6. The location of transmission zeros, a numerical example

In this section we give a numerical example which is taken from an important application of the previous theory viz. when $F(x)$ is an autocorrelation function for a discrete time stationary stochastic process. This is only a special case of what is developed in the previous sections but, because of the symmetry in the problem, it simplifies the formulae and shows clearly the applicability of the theory. So e.g. $f_k = \bar{f}_{-k}$ and $F(x)$ is positive real on the unit circle. Also $\alpha_i = \beta_i$ $\forall i$ so that $\mathcal{L}_n^\alpha = \mathcal{L}_n^\beta = \mathcal{L}_n$.

The general theory for this application has appeared elsewhere [2,3,7,8,9] and it is not repeated here. We mention only the main results.

The most important consequence of the specialisation is that we can now give a least squares optimality interpretation for the approximant, weighted with $F(x)$ itself. Indeed, suppose $F(x)$ has the factorization $s(x).\overline{s(1/\bar{x})}$, then

$$\min_{h_n \in \mathcal{L}_n} \frac{1}{2\pi} \int_{-\pi}^{\pi} \left|\frac{1}{s(e^{j\theta})} - h_n(e^{j\theta})\right|^2 F(e^{j\theta})d\theta$$

is attained for $h_n = k_n(x,0)s(0)$ and the minimum is given by $S_n = 1 - k_n(0,0)s(0)^2$. Remark however that the solution is parametrized in the chosen transmission zeros α_i that define $\mathcal{L}_n$. So the previous methods do not only give us an algorithm to find an approximant with given transmission zeros, but we can even improve upon this result if we are willing to optimize the least squares error S_n as a function of $\alpha_0, \alpha_1, \alpha_2, \ldots$. Therefore we have to maximize $k_n(0,0)$, or equivalently to minimize $\prod_1^n (1-|\rho_i|^2)/(1-|\gamma_i|^2)$ (see corollary 5.2 with $\rho_i^\alpha = \rho_i^\beta = \rho_i$ and $\gamma_i = \alpha_i\rho_i$). Thus we have a tool to choose the optimal location of the interpolation points α_i (= transmission zeros). We shall even find the exact transmission zeros, provided the given $F(x)$ is rational and that the degree of the rational approximant is high enough (at least as high as the degree of the given function). The objective function to be minimized is evaluated after n steps of the recursive algorithm defined in section 3, but, taking the symmetry into account, it becomes even simpler. It is familiar with the interpolation algorithm of Nevanlinna and Pick [1]. So if n is not too high,a function evaluation is relatively cheap.

For a numerical example we took for $F(x)$ the Laurent series valid at $|x|=1$ for the function

$$\frac{(x-\alpha)^2}{(x-\beta_1)^2(x-\beta_2)(x-\bar{\beta}_2)} ; \quad \alpha = 0.5; \ \beta_1 = -0.5; \ \beta_2 = 0.5 \exp(j\pi/4)$$

The first coefficients are listed below :

$f_0 = 3.925130$	$f_1 = -3.205387$
$f_2 = 2.104909$	$f_3 = -1.367445$
$f_4 = 0.825844$	$f_5 = -0.478882$
$f_6 = 0.279871$	$f_7 = -0.156157$
$f_8 = 0.087150$	$f_9 = -0.048429$
$f_{10} = 0.026176$	$f_{11} = -0.014318$
$f_{12} = 0.007713$	$f_{13} = -0.004113$
$f_{14} = 0.002214$	$f_{15} = -0.001170$
$f_{16} = 0.000610$	$f_{17} = -0.000328$
$f_{18} = 0.000171$	$f_{19} = -0.000090$

If we estimate the zeros at the origin and at $\alpha = 0.5$ exactly, then we find after four steps of the Schur algorithm the exact approximant with a minimal value of $S_4 = 0.90398$. However numerical experiments show that the location of α at 0.5 is not crucial at all because any other value of α in the interval [0.4,0.8] (if we know α to be real) gives about the same minimum. This illustrates that the approximation is in many cases (even in this simple example) rather insensitive to an exact location of the transmission zeros. If however we look at this procedure as a method to find the exact transmission zeros, then the example illustrates the ill conditioning of this problem. It is comparable with a situation of least squares exponential approximation or any other least squares rational approximation problem. Other, more complicated examples all showed the same characteristic behaviour, the situation becoming worse as the transmission zeros approached the unit circle.
The convergence of this special case has been studied at least theoretically.

The general situation as described in previous sections however is still an open question. In this case we can also have non-normal situations as for the Padé approximation problem. The study of these anomalies is still under investigation and will be reported on later. The theory is however appealing and several potential applications in minimax rational approximation [10], systems theory, coding theory [13], networks [9] and scattering theory [12] are developed.

References

1. Akhiezer, N.I. :The classical moment problem, Oliver and Boyd, Edinburgh-London, 1965.

2. Bultheel, A. : Recursieve rationale benaderingen; Ph.D. Thesis, K.U.Leuven, Sept. 1979.

3. Bultheel, A. , Dewilde, P. : On the optimal location of transmission zeros in least squares ARMA filtering, Report TW 51, K.U.Leuven, June 1980.

4. Bultheel, A., Dewilde, P. : On the relation between Padé approximation and Levinson/Schur recursive methods in M. Kunt - F. de Coulon (eds.), Proceedings EUSIPCO-80, North-Holland, Amsterdam, 1980, 517-523.

5. Bognar, J. : Indefinite inner product spaces, Springer-Verlag, Berlin,1974.

6. Chisholm, J.S.R., Common, A. : Chebyshev Padé approximants, in L. Wuytack (ed.), Padé approximation and its applications, Springer-Verlag, Berlin, (1979), 1-19.

7. Dewilde, P., Dym, H. : Schur recursions, error formulas and convergence of rational estimators for stationary stochastic sequences, to appear in IEEE Trans. Inf. Theory.

8. Dewilde, P. : On the convergence of the generalized Szegö-Levinson least square error algorithm, Tech. Rept. 82, TH Delft, 1979.

9. Dewilde, P., Vieira, A., Kailath, T. : On a generalized Szegö-Levinson realization algorithm for optimal linear predictors based on a network synthesis approach, IEEE Trans. CAS-25, (1978), 663-675.

10. Genin, Y., Kung, S.Y. : A two-variable approach to the model reduction problem with Hankel norm criterion. Subm. for publication IEEE Trans. 1980.

11. Gragg, W.B. : Laurent, Fourier and Chebyshev-Padé tables in [15] (1977), 61-72.

12. Lax, S.D., Phillips, R.S. : Scattering theory, Academic Press, New York, 1967.

13. McEliece, R.J. : The theory of information and coding, Addison-Wesley, Reading, Mass., 1977.

14. Meschkowski, H. : Hilbertsche Räume mit Kernfunktion, Springer-Verlag, Berlin, 1962.

15. Saff, E.B., Varga, R.S. (eds.) : Padé and rational approximation, Academic Press, New York, 1977.

16. Schur, I. : Über Potenzreihen die im Innern des Einheitskreises beschränkt sind, Z. Reine Angew. Math. 147 (1917), 205-232.

17. Thron, W.J. : Two-point Padé tables, T-fractions and sequences of Schur, in [15], (1977), 215-226.

18. Wiener, N., Masani, P. : The prediction theory of multivariate stochastic processes, Acta Mathematica, 98 (1957), 111-150; 99 (1958), 93-139.

Adhemar BULTHEEL
Katholieke Universiteit Leuven
Afdeling Toegepaste Wiskunde en Programmatie
Celestijnenlaan 200 A
B-3030 Heverlee (BELGIUM)

APPENDIX :

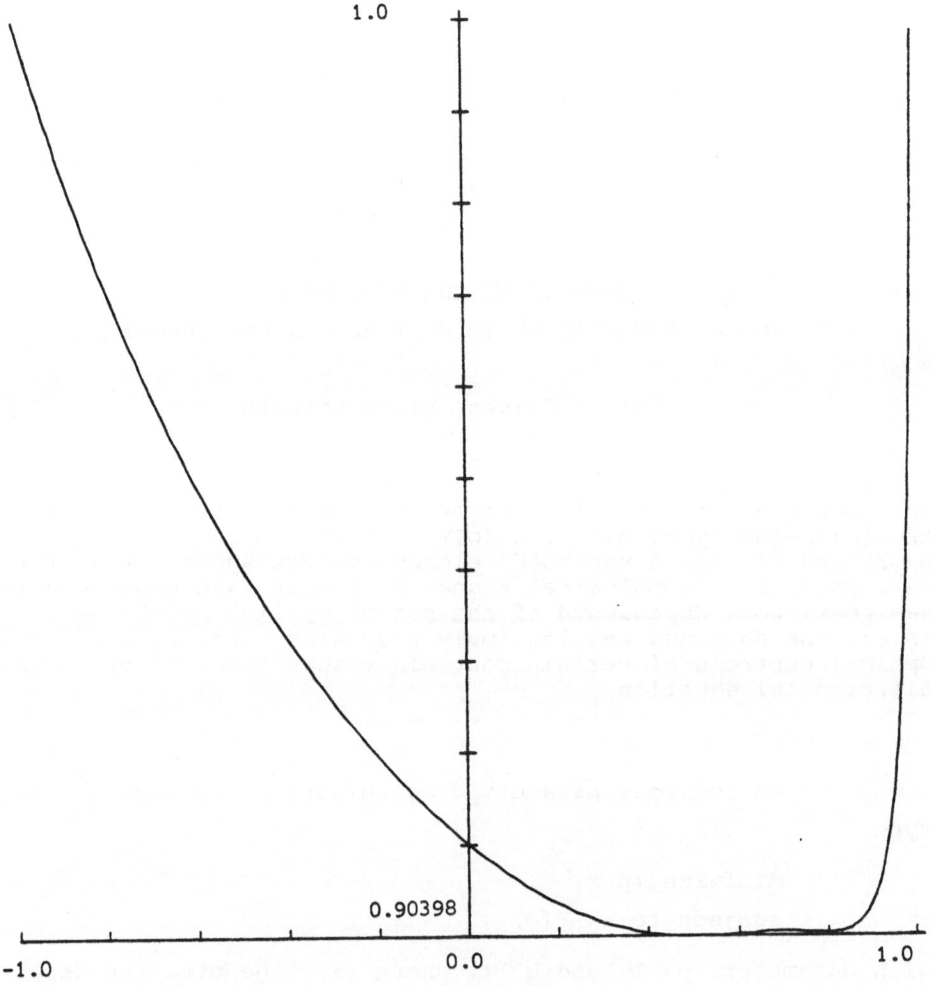

Objective function $\prod_{1}^{4}(1-|\rho_i|^2)/(1-|\gamma_i|^2)$ for the numerical example after two extractions at the origin and two extractions in $\alpha \in [-1,1]$.

The minimum 0.90398 is obtained for $\alpha = 0.5$.

PARAMETRIC OPTIMIZATION AND AN APPLICATION TO OPTIMAL CONTROL THEORY

Rainer Colgen, Klaus Schnatz

This paper treats stability theory for optimization problems of the following type: Minimize (p,x) subject to $x \in F(\varphi)$, where $p \in \mathbb{R}^m$ and $F(\varphi)$ is a non-void, closed, convex subset of $\mathbb{R}^m$ for $\varphi \in \Phi$ which is a topological space. We investigate upper and lower semicontinuous dependence of the set of optimal solutions on (p,φ). The obtained results imply piecewise continuity of time-optimal controls of certain control systems governed by ordinary differential equations.

We consider parametric optimization problems of the type

Minimize (p,x)
subject to $x \in F(\varphi)$

with parameters $p \in \mathbb{R}^m$ and $\varphi \in \Phi$, where $(\cdot,\cdot)$ denotes the usual scalar product in $\mathbb{R}^m$ and $F(\varphi)$ is a non-void, closed, convex subset of $\mathbb{R}^m$ for all $\varphi \in \Phi$ which is a topological space.

We first state some definitions: We call Φ the set of feasible parameters and $F(\varphi)$ the set of feasible points for all $\varphi \in \Phi$. Furthermore, we define the optimal value

$$E(p,\varphi) := \inf \{(p,x) : x \in F(\varphi)\}$$

and the set of optimal solutions

$$S(p,\varphi) := \{x \in F(\varphi) : (p,x) = E(p,\varphi)\}$$

for all $(p,\varphi) \in \mathbb{R}^m \times \Phi$ and the solvability set

$$\Sigma := \{(p,\varphi) \in \mathbb{R}^m \times \Phi : S(p,\varphi) \neq \emptyset\}.$$

In the sequel, we investigate upper and lower semicontinuity of the mapping

$$S: \quad \Sigma \longrightarrow POT(\mathbb{R}^m)$$
$$(p,\varphi) \longrightarrow S(p,\varphi).$$

During the investigation of upper semicontinuity of S, the mapping

$$F: \Phi \longrightarrow POT(\mathbb{R}^m)$$
$$\varphi \longrightarrow F(\varphi)$$

plays an important role, as is seen in the next theorem.

<u>THEOREM 1:</u> <u>Let</u> $(p,\varphi) \in \Sigma$ <u>such that</u> $S(p,\varphi)$ <u>is compact and</u> $0 \in F(\psi)$ <u>for all</u> $\psi \in \Psi$, <u>a neighborhood of</u> φ <u>in</u> Φ; <u>assume that</u> F <u>is a closed mapping</u>. <u>Then</u> S <u>is upper semicontinuous in</u> (p,φ).

<u>Proof:</u> Let $((p_n,\varphi_n))$ be a sequence in Σ with $(p_n,\varphi_n) \longrightarrow (p,\varphi)$.

We first show that every sequence (x_n) in $\mathbb{R}^m$ with $x_n \in S(p_n,\varphi_n)$ for all $n \in \mathbb{N}$ is bounded:

Suppose that (x_n) is unbounded; without loss of generality, we may then assume that $\|x_n\| \longrightarrow \infty$. $(x_n/\|x_n\|)$ is a bounded sequence and thus has a convergent subsequence; without loss of generality, we may assume that

$$x_n/\|x_n\| \longrightarrow x \in \mathbb{R}^m.$$

Since $0 \in F(\varphi_n)$ and $x_n \in S(p_n,\varphi_n) \subset F(\varphi_n)$ for all $n \in \mathbb{N}$, we obtain that for every $\lambda \geq 0$ exists $n_0 \in \mathbb{N}$ such that $(\lambda/\|x_n\| \leq 1$ and thus) $\lambda x_n/\|x_n\| \in F(\varphi_n)$ for all $n \in \mathbb{N}$ with $n \geq n_0$. As F is a closed mapping, we conclude that $\lambda x \in F(\varphi)$ for all $\lambda \geq 0$ which means that x belongs to the characteristic cone of $F(\varphi)$. (Recall that the characteristic cone of a closed, convex set C is defined as $\{z \in \mathbb{R}^m : y+\lambda z \in C$ for all $y \in C$ and $\lambda \geq 0\}$; see [3].) From $0 \in F(\varphi_n)$ we obtain that $E(p_n,\varphi_n) \leq 0$ for all $n \in \mathbb{N}$ sufficiently large and thus

$$(p,x) = \lim (p_n, x_n/\| x_n \|) = \lim E(p_n,\varphi_n)/\| x_n \| \leq 0.$$

Since $(p,\varphi) \in \Sigma$ and x belongs to the characteristic cone of $F(\varphi)$, an easy consideration provides that $(p,x) \geq 0$ and thus $(p,x) = 0$. So x belongs to the characteristic cone of $S(p,\varphi)$, too, which establishes a contradiction to the compactness of the set of optimal values for (p,φ).

Now upper semicontinuity of S in (p,φ) is proved by standard arguments.

In the sequel, we treat lower semicontinuity of S. For the sake of simplicity, we only consider the case that $(p,\varphi) \in \mathrm{int}(\Sigma)$, the topological interior of Σ.

THEOREM 2: *Let* $(p,\varphi) \in \mathrm{int}(\Sigma)$. *Then* S *is lower semicontinuous in* (p,φ) *iff* $S(p,\varphi)$ *is a singleton.*

Theorem 2 is an easy corollary of the next lemma:

LEMMA: *Let* $(p,\varphi) \in \Sigma$ *such that* $S(p,\varphi)$ *has two extremal points* x_1 *and* x_2.

i) *If* (p_n,φ) *is a sequence in* Σ *such that* $(p_n, x_2-x_1) > 0$ *for all* $n \in \mathbb{N}$, *then* $x_2 \notin S(p_n,\varphi)$ *for all* $n \in \mathbb{N}$.

ii) *If* (p_n,φ) *is a sequence in* Σ *such that there is a zero sequence* (λ_n) *with* $p_n = p + \lambda_n(x_2-x_1)$ *for all* $n \in \mathbb{N}$, *then* x_2 *is no accumulation point of optimal solutions for* (p_n,φ).

REMARK: The preceding results especially hold for semi-infinite linear optimization problems where $F(\varphi)$ is given by

$$F(\varphi) := \{x \in \mathbb{R}^m : ((a(\varphi))(t), x) \leq (b(\varphi))(t) \text{ for all } t \in T\},$$

where T is a compact Hausdorff space and the mapping from the set of feasible parameters into $C(T;\mathbb{R}^m) \times C(T)$ which assigns to every $\varphi \in \Phi$ the pair $(a(\varphi), b(\varphi))$ is continuous.

In this case, Theorems 1 and 2 may be applied if the set of optimal solutions is compact and the set of feasible points satisfies a Slater-condition which means that there exists $x \in F(\varphi)$ such that

$$((a(\varphi))(t),x) < (b(\varphi))(t) \quad \text{for all } t \in T,$$

for a given parameter in the solvability set.

Theorem 1 then yields a generalization of a result of [1] where it is presumed that for every sequence (p_n,a_n,b_n) in the solvability set (which here is a subset of the parameter space $\mathbb{R}^m \times C(T;\mathbb{R}^m) \times C(T))$ converging to (p,a,b), the given parameter in the solvability set, there is a bounded selection, i.e., a bounded sequence (x_n) in $\mathbb{R}^m$ with x_n being an optimal solution for (p_n,a_n,b_n).

We finish this paper by presenting an application to optimal control theory which slightly generalizes some results of [2] and serves as preparatory result for numerical calculations, too.

<u>THEOREM 3: Let</u> u <u>be an optimal control for the problem</u>

<u>Minimize</u> T

<u>subject to</u> $\dot{z}(t) = Az(t) + Bu(t)$ <u>for all</u> $t \in [0,T]$;

$z(0) \in \mathbb{R}^n$; $u(t) \in F$ <u>for all</u> $t \in [0,T]$,

<u>where</u> $A \in \mathbb{R}^{n,n}$, $B \in \mathbb{R}^{n,m}$ <u>and</u> F <u>is a non-void, compact, convex subset of</u> $\mathbb{R}^m$ <u>with finitely many extremal subsets. Further assume that for every</u> $w \in \mathbb{R}^m$ <u>parallel to any extremal subset of</u> F <u>holds</u>

$$\dim(\operatorname{span}(Bw,ABw,\ldots,A^{n-1}Bw)) = n.$$

<u>Then</u> u <u>is piecewise continuous.</u>

The <u>proof</u> of Theorem 3 is performed by the aid of our preceding results, the maximum principle [2] which states that there is a solution of the ordinary differential equation

$$\dot{\lambda}(t) = -A^T\lambda(t) \quad \text{for all } t \in [0,T]$$

such that u(t) is a solution of the optimization problem

Maximize $(B^T\lambda(t),u)$

subject to $u \in F$

for all $t \in [0,T]$ and the fact that there are only finitely many $t \in [0,T]$ such that the corresponding optimization problem is not uniquely solvable because F has only finitely many extremal subsets besides those consisting of one extremal point, λ is an analytic function and $\dim(\mathrm{span}(Bw, ABw, \ldots, A^{n-1}Bw)) = n$ for all $w \in \mathbb{R}^m$ parallel to any of these extremal subsets of F.

REFERENCES:

1. Brosowski,B.: Parametric approximation and optimization. Amsterdam, North Holland, to appear.

2. Damm,H; Kopielski,T.: Einige Anwendungen der linearen parametrischen Optimierung bei der Untersuchung von optimalen Steuerungsproblemen; in: Lommatzsch,K.: Anwendungen der linearen parametrischen Optimierung. Basel-Stuttgart, Birkhäuser 1979.

3. Grünbaum,B.: Convex polytopes. London-New York-Sydney, Interscience Publishers 1967.

Rainer Colgen, Klaus Schnatz
Fachbereich Mathematik der Universität Frankfurt/Main
Robert-Mayer-Str. 6 - 10
D - 6000 Frankfurt/Main - 1
Federal Republic of Germany

NUMERICAL ALGORITHMS FOR LEAST SQUARES APPROXIMATION BY MULTIVARIATE B-SPLINES

Wolfgang A. Dahmen and Charles A. Micchelli

This paper is concerned with some computational aspects of the two basic ingredients for Galerkin-type approximations by linear combinations of multivariate B-splines, namely the numerical evaluation of the corresponding inner products and the practical construction of a well-conditioned spline basis. The respective algorithms proposed in this paper are based on some recent results which are briefly surveyed on occasion. The discussion of the algorithms is supplemented by numerical examples.

1. INTRODUCTION.

Various recent studies enhance the pivotal position of the multivariate B-spline in a new approach towards globally smooth spline approximation [5-13,15-17,20,21]. The uniformity of the purely 'knot dependent' structure for arbitrary spatial dimensions and any degree as well as the extreme flexibility of the B-splines are not only very attractive from a theoretical point of view but may also bear a considerable potential for practical applications.

However the characteristics of the B-splines do not seem to fit very well into the traditional multivariate numerical concepts. So, to begin with discussing possibly simple numerical examples involving B-splines may facilitate taking finally optimal practical advantage of such a knot dependent concept.

The primary purpose of this paper is therefore to report on our experiences with least squares approximation by multivariate B-splines. This should be viewed however as one example of Galerkin-type approximations suggesting further applications e.g. as 'conforming Finite Element' schemes for higher order problems. Our approach is essentially based on some recent results [9-11] which we wish to survey on occasion.

After briefly summarizing in section 2 some known properties of multivariate B-splines for later reference, we will discuss in section 3 several approaches to the central task of computing inner products of multivariate B-splines. In particular, approximative formulas are compared with exact algorithms which are based on certain recurrence relations [9,10] exhibiting the same structure as the known recursions for point evaluations [5, 15,16,20]. Section 4 deals with the practical construction of 'well-conditioned' spline basis. Elaborating on some specifications of a more general construction principle [9,11] leads to the realization of a smooth 2-variate least squares scheme. Some numerical tests are reported in section 5.

2. SOME PROPERTIES OF MULTIVARIATE B-SPLINES

In this section we review a few known properties of multivariate B-splines for later use.

To this end, let us start with fixing some notation. Elements of the Euclidean space R^s are typically denoted by x, z, $x = (x_1,\dots,x_s)$. The (closed) convex hull of a given set A in R^s, its s-dimensional Lebesgue measure and its characteristic function are denoted by $[A]$, $\mathrm{vol}_s(A)$ and $\chi_A(x)$, respectively, whereas $|A|$ means the cardinality of any finite set A.

Adopting the standard multiindex notation for $\alpha, \beta \in Z_+^s$, i.e. $|\alpha| = \alpha_1+\dots+\alpha_s$, $x^\alpha = x_1^{\alpha_1},\dots,x_s^{\alpha_s}$,

$$\Pi_{s,k} = \{ \sum_{|\alpha|\leq k} c_\alpha x^\alpha : x \in R^s, c_\alpha \in R \}$$

is the space of s-variate polynomials of (total) degree $\leq$ k.

For any knot set $K = \{x^0, \dots, x^n\} \subset R^s$ which will be always assumed to satisfy

(2.1) $$\mathrm{vol}_s([K]) > 0$$

the corresponding B-spline $M(x|K) = M(x|x^0, \dots, x^n)$ is conveniently defined by requiring that [20]

(2.2) $$\int_{R^s} M(x|x^0, \dots, x^n) f(x)dx = n! \int_{S^n} f(t_0 x^0 + \dots t_n x^n)dt_1 \dots dt_n$$

holds for any continuous function f. Here

$$S^n = \{(t_1, \dots, t_n) : \sum_{j=1}^{n} t_j = 1-t_0,\ t_j \geq 0,\ j=0, \dots, n\}$$

is the standard n-simplex. While (2.2) reveals that $M(x|K)$ has compact support $[K]$ the following recurrence relation due to the second author [20] (cf. also [5, 15, 16, 21]) affirms that it is a piecewise polynomial of degree n-s:

(2.3) $$M(x|x^0, \dots, x^n) = \frac{n}{n-s} \sum_{j=0}^{n} \lambda_j M(x|x^0, \dots, x^{j-1}, x^{j+1}, \dots, x^n)$$

whenever $n > s$ and $x = \sum_{j=0}^{n} \lambda_j x^j$, $\sum_{j=0}^{n} \lambda_j = 1$. For $n = s$ one has

(2.4) $$M(x|x^0, \dots, x^s) = (\mathrm{vol}_s([x^0, \dots, x^s]))^{-1} \chi_{[x^0, \dots, x^s]}(x).$$

Of course, (2.3), (2.4) admit a recursive numerical evaluation of the B-spline. Concerning a more detailed discussion of corresponding algorithms and their stability properties the reader is referred to [7, 21].

A similar relation holds for the derivatives of the B-spline. Setting $D_z f = \sum_{1 \leq i \leq s} z_i \frac{\partial f}{\partial x_i}$ one has [5, 20]

(2.5) $$D_z M(x|x^0, \dots, x^n) = n \sum_{o \leq j \leq n} \mu_j M(x|x^0, \dots, x^{j-1}, x^{j+1}, \dots, x^n)$$

whenever $z = \sum_{j=0}^{n} \mu_j x^j, \quad \sum_{j=0}^{n} \mu_j = 0$ and the right hand side of (2.5) is defined in x.

One may use (2.5) to show that $M(x|K)$ possesses continuous derivatives of order n-s-d if the convex hull of any s+d knots in K has dimension s. In particular

(2.6) $M(x|x^0,\ldots,x^n) \in C^{n-s-1}(R^s)$ if the x^i are in 'general position',

i.e. if any s+1 of the knots are affinely independent (cf. [5, 20]) Recently, Hakopian [15] showed that the above conditions are sharp.

Further functionals of practical importance are certainly those involving integration. To this end, an alternate representation of the B-spline turns out to be useful which is based on a notion of multivariate truncated powers [5] (cf. also [9, 21]). For any $x^1,\ldots,x^m \in R^s$ such that

(2.7) $$0 \notin [x^1,\ldots,x^m]$$

we define $G(x|x^1,\ldots,x^m)$ by requiring that

(2.8) $$\int_{R^s} f(x)G(x|x^1,\ldots,x^m)\,dx = \int_0^\infty \cdots \int_0^\infty f(t_1x^1+\ldots+t_mx^m)\,dt_1\ldots dt_m$$

holds for any locally supported continuous function f. Choosing for arbitrary $0 \neq x^1,\ldots,x^m \in R^s$ some fixed 'orientation' $\varepsilon = \{\varepsilon_1,\ldots,\varepsilon_m\}$, $\varepsilon_i \in \{-1,1\}$, such that the $\varepsilon_i x^i$ satisfy (2.7) $H = H_\varepsilon$ is given by [5]

$$H(x|x^{i_1},\ldots,x^{i_q}) = \Big(\prod_{j=1}^{q} \varepsilon_{i_j}\Big)G(x|\varepsilon_{i_1}x^{i_1},\ldots,\varepsilon_{i_q}x^{i_q})$$

for any $1 \leq i_1<\ldots< i_q \leq m$, $1 \leq q \leq m$.

In analogy to the univariate case truncated powers and B-splines are interrelated as follows. Let $H = H_\varepsilon$ be defined with respect to some fixed orientation of the set $\{x^i - x^j : 0 \leq j<$

$< i \leq n\}$ when $x^0,\dots,x^n$ are pairwise distinct knots, then

$$(2.9)\quad M(x|x^0,\dots,x^n) = n! \sum_{j=0}^{n} (-1)^j H(x-x^j|x^j-x^0,\dots,x^j-x^{j-1},x^{j+1}-x^j,\dots,x^n-x^j).$$

For similar representations in the case of coalescent knots see [6]. We shall have to use later some more properties of the truncated powers. Assuming that the $x^1,\dots,x^s$ span R^s one has [5]

$$(2.10)\quad G(x|x^1,\dots,x^s) = (s!\,\mathrm{vol}_s([0,x^1,\dots,x^s]))^{-1}\chi_{<x^1,\dots,x^s>_+}(x)$$

where $<x^1,\dots,x^s>_+ = \{\sum_{1\leq j\leq s} t_j x^j : t_j \geq 0,\ j=1,\dots,s\}$.

In general $G(x|x^1,\dots,x^m)$ is a piecewise polynomial of degree m-s. Specifically, higher order truncated powers can be obtained by convolving lower order ones, i.e. [5]

$$(2.11)\quad H(x|x^1,\dots,x^m) = (H(\cdot|x^1,\dots,x^q) * H(\cdot|x^{q+1},\dots,x^m))(x)$$

(where the convolution is to be understood eventually in the distributional sense).

Aside from (2.11) there is also an 'algebraic' recursion similar to (2.3) (cf. [5, 9, 21])

$$(2.12)\quad H(x|x^1,\dots,x^m) = \frac{1}{m-s}\sum_{j=1}^{m} \lambda_j H(x|x^1,\dots,x^{j-1},x^{j+1},\dots,x^m)$$

when $m > s$ and $x = \sum_{j=1}^{m} \lambda_j x^j$.

3. INNER PRODUCTS OF MULTIVARIATE B-SPLINES

An essential precondition for applying multivariate B-splines in a least squares or Finite Element setting is the efficient numerical evaluation of quantities like

(3.1) $$\int_{R^s} f(x)M(x|x^o,\dots,x^n)dx$$

as a typical 'right hand side' component and

(3.2) $$\int_{R^s} DM(x|x^o,\dots,x^n)\ DM(x|y^o,\dots,y^m)dx$$

say, as an entry of the associated 'stiffness matrix' where D may stand for a linear differential operator. However, note that (3.2) may ultimately be reduced to expressions of the following type

(3.3) $$I(x^o,\dots,x^n|y^o,\dots,y^m) = \int_{R^s} M(x|x^o,\dots,x^n)M(x|y^o,\dots,y^m)dx$$

which readily follows from (2.5).

To begin with the discussion of the general quantities (3.1) the direct application of a single cubature formula seems unreasonable. In fact, the actual integration domain is hard to classify for arbitrary knots and the approximation would depend on the smoothness of $M(x|K)$ which belongs at most to $C^{n-s-1}(R^s)$. On the other hand there is in general no practicable way to decompose $[K]$ into those subdomains where $M(x|K)$ is polynomial.

An alternate approach, namely to apply an n-dimensional cubature formula to the right hand side of (2.2), exploits the original geometrical interpretation of the B-spline [2]. To this end, suppose that for $g: R^n \to R$

(3.4) $$\sum_{i=1}^{N} a_i\ g(u^i) \cong \int_{S^n} g(u)\ du$$

is a cubature formula of degree d for the standard n-simplex S^n. Rewriting the right hand side of (2.2) as

$$\int_o^1 \dots \int_o^{u_{n-1}} f(x^o+u_1(x^1-x^o)+u_2(x^2-x^o)+\dots+u_n(x^n-x^o))du_n\dots du_1$$

and setting $g(u) = f(x^o+u_1(x^1-x^o)+\dots+u_n(x^n-x^o))$ (3.4) provides the approximation

(3.5) $$\int_{R^s} M(x|x^o,\dots,x^n) f(x)dx \cong n! \sum_{i=1}^{N} a_i\, f(x^o + Hu^i)$$

which is exact for $f \in \Pi_{s,d}$. Here, u^i are the original integration nodes for S^n and H is the sxn matrix with columns $x^i - x^o$, $i=1,\dots,n$.

Of course, (3.5) is essentially an n-variate formula and requires therefore more function evaluations than an s-variate one. However, since in most practical cases n will not exceed s by more than three, say, this drawback is certainly compensated for by some obvious advantages: No prior evaluations of the B-spline are needed. The application of (3.5) is simple and works for arbitrary knot positions.

(3.5) will be used in the subsequent section for the computation of the right hand side components of normal equations. Let us briefly illustrate here its performance with respect to the behavior of $M(x|K)$ as an 'approximate identity'.

Let $f(x) = \cos(7(x+y))+0.3$ and $K = \{(0,0),(0.6,0.76),(0.9, 0.82), (1,1), (1,0)\}$. For scaling factors $m = 1, 5, 10, 100, 1000$ (3.5) combined with a fifth-order formula for the 4-simplex (cf. [24, p. 312]) yields approximations A_m for $\int_{R^2} f(x)M(x|K/m)dx$ shown in the first column of the following table. These values may be compared with $f(a^m)$ in the second column where a^m is the average of the knots in K/m.

m	A_m	$f(a^m)$
5	0.1211962	0.1687759
10	0.8958227	0.9590811
100	1.2954987	1.2963795
1000	1.2999549	1.2999638

Of course, the quantities $I(x^o,\dots,x^n|y^o,\dots,y^m)$ in (3.3) may now be viewed just as special cases of (3.1) where f happens to be also a B-spline. However, the above approach would not work very well in this case since the approximation would depend again on the smoothness properties of the B-spline which plays the role of f in (3.5). Moreover, the relative error of such an approximation will essentially not decrease when turning to 'finer mesh-sizes', e.g. by rescaling a given configuration of knot sets. Indeed, recalling that [21] $M(x|Ax^o,\ldots,Ax^n) = |\det A|^{-1} M(A^{-1}x|x^o,\ldots,x^n)$ holds for any linear transformation $A: R^s \to R^s$ we obtain

$$
(3.6)\qquad \begin{aligned} I(AK|AK') &= (\det A)^{-2} \int_{R^s} M(A^{-1}x|K)M(A^{-1}x|K')dx \\ &= |\det A|^{-1} I(K|K'). \end{aligned}
$$

On the other hand, (3.5) behaves similarly

$$
n! \sum_{i=1}^{N} a_i M(Ax^o+AHu^i|AK') = |\det A|^{-1} n! \sum_{i=1}^{N} a_i M(x^o+Hu^i|K').
$$

The above negative predictions are confirmed by some numerical experiments reported below. Consider the 2-variate knot sets

$K_1 = \{(0,0),\ (0.6,0.76),\ (0.9,0.82),\ (1,1),\ (1,0)\}$,

$K_2 = \{(0,0),\ (0.6,0.76),\ (-0.1,-0.18),\ (-0.4,-0.24),\ (-0.6,-0.24)\}$,

$K_3 = \{(0,0),\ (1,0),\ (0.9,0.82),\ (0.6,0.76),\ (0.9,-0.18)\}$,

$K_4 = \{(0,0),\ (0.6,0.76),\ (-0.1,-0.18),\ (-0.1,0.82),\ (-0.4,0.76)\}$,

which correspond to B-splines of degree two in the plane. In order to approximate $I(K_1|K_i)$ by means of (3.5) we have used a third and a fifth degree formula with non-negative weights and 15 and 31 nodes, respectively (cf. [24, pp. 308, 312]). The two values associated with each pairing K_1, K_i in the first two columns of the following table are the different results which are obtained when interchanging the roles of the B-splines in (3.5). The last column contains the exact (up to round off) values of the respective inner products which were computed by means of certain recurrence relations discussed later in this section.

Aside from the fact that even for the fifth-degree formula the relative errors are still in the range of hundred percent we see that the approximative results depend significantly on which B-spline plays the role of f in (3.5).

	d = 3	d = 5	$I(K_1 \mid K_i)$
K_1, K_2	0.1578749 0.2761607	0.1185027 0.1651156	0.1233675
K_1, K_3	2.0399539 1.8228357	1.9870161 1.9452133	1.1847322
K_1, K_4	0.0023839 0.0021500	0. 0.	0.0000156
K_1, K_1	4.2962638	3.7850684	2.7735750

Let us briefly mention an alternate possibility of approximately calculating the inner products $I(K|K')$ using the Fourier transform $(M(\cdot|K))^{\wedge}(u)$ of $M(x|K)$. To this end, let $x \cdot z$ denote the standard scalar product on R^s so that $\|x\| = (x \cdot x)^{1/2}$ is the Euclidean norm of $x \in R^s$. Using the Hermite-Genocchi formula for divided differences one may derive from (2.2)

$$(3.7) \qquad \begin{aligned} (M(\cdot|x^o,\dots,x^n))^{\wedge}(u) &= (M(\cdot|\tilde{u}\, x^o,\dots,\tilde{u}\cdot x^n))^{\wedge}(t) \\ &= [-i\tilde{u}\cdot x^o,\dots,-i\tilde{u}\cdot x^n]e^{\cdot t} \end{aligned}$$

where $t = \|u\|$, $\tilde{u} = u/t$. Hence Plancherel's formula yields

$$I(x^o,\dots,x^n|y^o,\dots,y^m)$$

$$= \int_{\|\lambda\|=1} \int_o^{\infty} (M(\cdot|\lambda\cdot x^o,\dots,\lambda\cdot x^n))^{\wedge}(t)\,(M(\cdot|\lambda\cdot y^o,\dots,\lambda\cdot y^m))^{\wedge}(t)\,dt\,d\lambda .$$

Thus choosing an integration formula for the n-sphere one may try to evaluate the inner univariate integral for a discrete set of points λ^i on the unit ball by either approximately calculating first the Fourier transform of the corresponding univariate B-splines or exploiting the fact that these transforms are in view of (3.7) actually divided differences of the exponential function.

Instead of pursuing this aspect let us turn the attention now to the exact evaluation of $I(K|K')$ for arbitrary knot sets K, K'. The following discussion is based on some recent results [10] on recursive representations for $I(K|K')$.

In order to exploit the truncated power representation (2.9) of the B-spline we have to assume first that the knots x^i and y^i in $K = \{x^o,\dots,x^n\}$, $K' = \{y^o,\dots,y^m\}$ are pairwise distinct, respectively.

ALGORITHM A

Replacing the B-splines by their respective truncated power representation (2.9) and using their convolution structure (2.11) we obtain [10]

$$(3.8)\quad I(K|K') = n!m! \sum_{i=o}^{n} \sum_{j=o}^{m} (-1)^{i+j} H(y^j-x^i|y^o-y^j,\dots,y^j-y^m, x^i-x^o,\dots,x^n-x^i).$$

So the integration is reduced to point evaluations of truncated powers which in turn may be processed by the recursion (2.12). Note that a suitable orientation for $H(x|x^1,\dots,x^m)$ (recall the remarks subsequent to (2.8)) can be found by setting $\varepsilon_1=1$, $\varepsilon_j=1$ if $x^1\cdot x^j > 0$, $\varepsilon_j=-1$, otherwise.

However, roughly $(n+1)(m+1)s^{n+m-s-1}$ linear sxs systems have to be solved during the recursion and (3.8) works only for pairwise distinct knots. Moreover the relatively large number of alternating summands may cause stability problems. On the other hand, when $n = s = m$ these negative effects are not nearly so sensible and, in particular, the inner product will then vanish unless the knots are affinely independent and hence pairwise distinct.

A first step to avoid the disadvantages for the general case $n, m > s$ is to introduce the functions

$$(3.9)\quad Q(x|y^1,\dots,y^d|x^o,\dots,x^n) = \int_{R^s} H(x-y|y^1,\dots,y^d)M(y|x^o,\dots,x^n)\,dy$$

which are defined now for arbitrary (possibly coalescent) knots x^i and (appropriately oriented) $y^j \neq 0$. Of course, when the knots x^i are pairwise distinct (2.11) and (2.9) provide

$$Q(x|y^1,\dots,y^d|x^o,\dots,x^n)$$

$$(3.10)\qquad = n! \sum_{i=o}^{n} (-1)^i H(x-x^i|y^1,\dots,y^d,x^i-x^o,\dots,x^n-x^i)$$

which in turn resembles again very much the representation (2.9). In fact, choosing λ_j such that

$$x = \sum_{j=o}^{n} \lambda_j x^j, \quad \sum_{j=o}^{n} \lambda_j = 1 \quad \text{and hence} \quad x-x^i = \sum_{j=o}^{n} \lambda_j (x^j-x^i),$$

applying (2.12) to each summand of the right hand side of (3.10) yields

$$(3.11)\qquad \begin{aligned} &Q(x|y^1,\dots,y^d|x^o,\dots,x^n) \\ &= \frac{n}{n+d-s} \sum_{j=o}^{n} \lambda_j Q(x|y^1,\dots,y^d|x^o,\dots,x^{j-1},x^{j+1},\dots,x^n) \end{aligned}$$

generalizing (2.3). Note that (3.11) is the precise analog to the univariate formula (1.7) in [3]. Since the B-splines depend continuously on their knots (cf.(2.2)) the relation (3.11) holds in view of (3.9) also for coalescent knots x^i.

ALGORITHM B

Combining (2.9) and (3.9) we arrive at the formula

$$(3.12)\qquad I(K|K') = m! \sum_{j=o}^{m} (-1)^j Q(y^j|y^o-y^j,\dots,y^j-y^m|x^o,\dots,x^n).$$

This suggests to

- apply (3.11) to each summand on the right hand side of (3.12) until $n = s$, then,
- using (3.10), the remaining truncated powers for $n = s$ are evaluated as in algorithm A by means of (2.12).

The error analysis for the analogous B-spline recursion (2.3) (cf. [7]) confirms that the recursion (3.11) is very stable at least when the argument y^j is in the convex hull of the knots x^i. But still the elements of one knot set have to be pairwise distinct and the number of alternating summands depends on the number $m > s$ of these knots. Moreover, it is easy to check that the complexity of algorithm B even exceeds that of algorithm A.

Let us present now a third algorithm which does not involve any restrictions to the knots. It is based on the following recursive representation of the inner products which covers in particular the univariate formulas in Lemma 4.1 [3] .

Choosing μ_j, η_j such that

$$\sum_{j=0}^{n} \mu_j x^j = \sum_{j=0}^{m} \eta_j y^j, \quad \sum_{j=0}^{n} \mu_j = 1 = \sum_{j=0}^{m} \eta_j$$

one has for $n,m > s$ [10]

$$I(x^o,\ldots,x^n|y^o,\ldots,y^m)$$

$$(3.13) \quad = \frac{1}{n+m-s}\Big(n \sum_{j=0}^{n} \mu_j I(x^o,\ldots,x^{j-1},x^{j+1},\ldots,x^n|y^o,\ldots,y^m)$$

$$+ \quad m \sum_{j=0}^{m} \eta_j I(x^o,\ldots,x^n|y^o,\ldots,y^{j-1},y^{j+1},\ldots,y^m)\Big).$$

This gives rise to

ALGORITHM C

i) When $|K|$, $|K'| > s+1$ (3.13) reduces $I(K|K')$ to a combination of lower order inner products which either vanish or involve one knot set consisting of exactly $s+1$ affinely independent knots. Hence

ii) algorithm B appplies to each of these terms, i.e (3.12) can be used with $m = s$.

Note that no restrictions to the knot positions are required. Since the coefficients μ_j, η_j can be always chosen nonnegative when $[K] \cap [K'] \neq \emptyset$, i.e. $I(K|K') \neq 0$, and since at the stage where algorithm B and hence also A are called the above mentioned disadvantages of these algorithms are minimized, algorithm C is certainly the stablest version.

On the other hand, its complexity grows even much faster than that of algorithm A or B since, when insisting on nonnegative μ_j, η_j in (3.13), a reduction by one knot produces in general 2s+2 lower order inner products.

However, algorithm C exhibits the most promising potential for saving computational work by exploiting certain interrelations between the knot sets corresponding to elements of reasonable B-spline bases. In fact, it will be shown in the subsequent section that $[K]\cap[K']\neq\emptyset$ typically implies $K\cap K'\neq\emptyset$. Then (3.13) simplifies for any s to

$$I(z,x^1,\dots,x^n|z,y^1,\dots,y^m) = \frac{1}{n+m-s}\,(nI(x^1,\dots,x^n|z,y^1,\dots,y^m)$$

$$(3.14) \qquad\qquad + \quad mI(z,x^1,\dots,x^n|y^1,\dots,y^m)).$$

Let us illustrate this effect for the extreme case K = K', n=m=4:

$$I(x^0,\dots,x^4|x^0,\dots,x^4)$$

(3.14): $\downarrow$

$$I(x^0,\dots,x^4|x^1,\dots,x^4)$$

(3.14): $\swarrow \qquad \searrow$

$$I(x^1,\dots,x^4|x^1,\dots,x^4) \qquad I(x^0,\dots,x^4|x^2,x^3,x^4)$$

(3.14): $\downarrow \qquad\qquad \downarrow$

$$I(x^1,\dots,x^4|x^2,x^3,x^4) \longrightarrow \quad \text{algorithm B.}$$

Moreover, when two knot sets have several common knots the evaluation points $y^j - x^i$ (cf. (3.8)) will frequently match the generating directions of the corresponding truncated powers, so that (2.12) simplifies for any s to

$$H(x|cx,x^1,\dots,x^m) = (c/m+1-s)H(x|x^1,\dots,x^m).$$

On the other hand, when K and K' have only a few common knots, several lower order inner products will vanish since the convex hulls of the reduced knot sets do sometimes not intersect each other. It turned out that checking these intersections is cheaper than going through the complete recursion tree.

Our implementation of algorithm C exploits the above facts. Dispensing perhaps with optimal stability we have also usually set $\mu_j = \delta_{ij}$, some i. APL is very convenient for handling the recursive structure of the algorithms as well as for efficiently managing the knot sets. E.g. reordering knots or evaluating the frequently occurring barycentric coordinates amounts to simply calling fixed APL functions.

Summarizing we note that on one hand the 'exact' algorithms tend to become expensive. On the other hand, the fact that all the basic formulae for the evaluation of the B-spline, its derivatives and for the inner products exhibit essentially the same consistent knot dependent structure should be a considerable advantage for practical realizations.

4. WELL CONDITIONED SPLINE BASES.

One essential precondition for any practical application of multivariate B-splines is certainly the availability of a suitable B-spline basis. A first general construction principle was proposed in [11] major parts of which appeared already in [9]. It seems that only little later than [9] K. Höllig presented essentially the same approach in [17] which we became aware of after finishing the draft of this paper.

It is a primary objective of this section to survey some of the results in [11] in order to elaborate then on certain specifications of the general setting. The subsequent discussion of the practical aspects of such a setting will be supplemented by some numerical examples. Specifically, combining the basis construction with the results of the previous section leads to the practical realization of a multivariate least squares scheme.

As usual we set $L_p(\Omega) = \{f : \|f\|_p(\Omega) = (\int_\Omega |f(x)|^p dx)^{1/p} < \infty\}$ $W_p^k(\Omega) = \{f : D^\alpha f \in L_p(\Omega), |\alpha| \leq k\}$ where $D^\alpha f = \partial^{|\alpha|} f/(\partial x)^\alpha$ with the familiar interpretation for $p = \infty$.

It is useful to start with a few elementary combinatorial considerations. A collection T of simplices is called a triangulation of G if $G = \bigcup\{\sigma : \sigma \in T\}$ and $\sigma \cap \sigma'$ is (empty or) a lower dimensional common face of $\sigma, \sigma' \in T$.

Denoting by S_n the set of all permutations of $\{1,\dots,n\}$, one may associate with any $\pi \in S_n$ the n-simplex $\sigma_\pi = [v_\pi^0,\dots,v_\pi^n]$

with vertices

(4.1) $$v_\pi^0 = 0, \quad v_\pi^j = v_\pi^{j-1} + e^{\pi(j)}, \quad j=1,\ldots,n,$$

$(e^i)_j = \delta_{ij}$. One readily verifies that

(4.2) $$K_n = \{\sigma_\pi : \pi \in S_n\}$$

is a triangulation of the unit n-cube $[0,1]^n$ known as 'Kuhn's triangulation' (cf. [19]). Let $\bar{e}^j = e^1+\ldots+e^j$, $j > 0$, $\bar{e}^0 = 0$ and $\Sigma^d = [0,\bar{e}^1,\ldots,\bar{e}^d]$. As part of a more general result it was shown in [9,11] that the following set of permutations

(4.3) $$A_{s,k} = \{\pi \in S_n : \pi(i_1),\ldots,\pi(i_s) \le s,\ s<\pi(j_1),\ldots,\pi(j_k) \le n \text{ for } i_1<\ldots<i_s,\ j_1<\ldots<j_k \text{ implies } \pi(i_q)=q,\ \pi(j_q)=s+q\}$$

induces a triangulation

$$T_{s,k} = \{\sigma_\pi : \pi \in A_{s,k}\}$$

of $\Sigma^s \times \Sigma^k$ and

(4.4) $$|T_{s,k}| = |A_{s,k}| = \binom{s+k}{s}$$

where we will always assume $n = s + k$.

More generally, let for a given s-simplex $\rho=[u^0,\ldots,u^s]$ F denote the affine map satisfying $F(\bar{e}^i) = u^i$, $i=0,\ldots,s$, so that $\bar{F} = F \otimes \mathrm{id}_k$ takes $\Sigma^s \times \Sigma^k$ into $\rho \times \Sigma^k$, then

(4.5) $$T_\rho = \{\bar{F}(\sigma) : \sigma \in T_{s,k}\}$$

is of course a triangulation of $\rho \times \Sigma^k$. Note that T_ρ is determined only by the ordering of the vertices u^i of ρ. In fact, recalling (4.1), (4.3) one may associate with any $\pi \in A_{s,k}$ $n+1$ pairs $(q_r, \ell_r) \in N \times N$, where $q_0, \ell_0 = 0$ and $q_r = q_{r-1}+1$, $\ell_r = \ell_{r-1}+1$ iff $r = i_m, j_m$ (cf. (4.3)), respectively, $r=1,\ldots,n$.

Hence

(4.6)
$$\sigma = [(u^{q_0},\bar{e}^{\ell_0}),\ldots,(u^{q_n},\bar{e}^{\ell_n})] \in T_\rho \quad \text{iff}$$
$$(q_r,\ell_r) \not\leq (q_{r+1},\ell_{r+1}),\ r=0,\ldots,n-1,\ 0\leq q_r\leq s,\ 0\leq \ell_r\leq k.$$

This readily implies that for any (d+k)-face $\delta = [u^{i_0},\ldots,u^{i_d}] \times \Sigma^k$

$$T_{[u^{i_0},\ldots,u^{i_d}]} = \{\sigma \cap \delta : \sigma \in T_\rho\} \quad ,$$

i.e. T_ρ induces triangulations of the same type on its lower- dimensional faces. In particular, when any two s-simplices $\rho = [u^0,\ldots,u^s]$, $\rho' = [v^0,\ldots,v^s]$ share some d-face $\delta = [a^0,\ldots,a^d]$ and $u^{i_j} = a^j = v^{q_j}$, $j=0,\ldots,d$, $q_0<\ldots<q_d$, $i_0<\ldots<i_d$, then the triangulations of $\delta \times \Sigma^k$ which are induced by T_ρ, $T_{\rho'}$, respectively, match up.

Hence, when for a given set $U = \{u^i : i=1,\ldots,N\}$ of points in R^s

$$\Theta = \{\rho(I) = [u^{i_0},\ldots,u^{i_s}] : I = (i_0,\ldots,i_s) \in J \subset Z_+^{s+1}\}$$

is a triangulation of $\Omega \subset R^s$, then

(4.7)
$$T_\Theta = \{T_{\rho(I)} : I \in J\}$$

is a triangulation of $\Omega \times \Sigma^k$.

It will be useful for later purposes to consider the special case $\Omega = [0,1]^s$, $\Theta = K_s$ (cf.(4.2)). Referring to Johnson's result [18] S_n may be ordered in such a way that the (j+1)st permutation $\pi_{n,j+1}$ is obtained from $\pi_{n,j}$ by one transposition of adjacent marks. In view of (4.1), (4.2) this induces an ordering

(4.8)
$$\sigma_{n,j} = \sigma_{\pi_{n,j}}$$

for K_n [1,11]. We showed in [11] that

(4.9)
$$K_{s,k} = \{\sigma_{n,j} : j=1,\ldots,n!/k!\} \subset K_n$$

is a sequential triangulation of $[0,1]^s \times \Sigma^k$. So, stopping the algorithm for K_n proposed in [1] after $n!/k!$ steps provides a convenient practical realization of $K_{s,k}$ for any $s, k \in Z_+$. Note that for $s, k > 1$ $T_{s,k}$ is not sequential. In general, one has (cf.(4.3)

(4.10) $$K_{s,k} = \{\sigma_\pi : s<\pi(j_1),\dots,\pi(j_k) \text{ for } j_1<\dots<j_k \text{ implies}$$
$$\pi(j_q) = s+q,\ q=1,\dots,k\} .$$

So, for $\Omega = [0,1]^s$, $\Theta = K_s$ we have $T_\Theta = K_{s,k}$ (cf.(4.7)).

Returning to the general situation (4.7) we associate now with every $u^i \in U$ $k+1$ points $u^{i,m}$, $m=0,\dots,k$, $u^{i,0} = u^i$, say. Then each $\rho(I) \in \Theta$ induces a configuration of knot sets

$$C_{\rho(I)} = \{ \{u^{i_{j_0},m_0},\dots,u^{i_{j_n},m_n}\} : i_{j_q} \in I,\ (j_q,m_q) \text{ satisfy } (4.6)\} .$$

Note that in view of (4.4)

(4.11) $$|C_{\rho(I)}| = \binom{s+k}{s}.$$

Setting

(4.12) $$C = \bigcup \{ C_\rho : \rho \in \Theta \}$$

we define the spline space

(4.13) $$S(C) = \text{span}\, \{M(x|K) : K \in C \} .$$

Each $K \in C$ may be viewed as the set of projected vertices of some n-simplex. In fact, (4.6) affirms that when $u^i = u^{i,m}$, $m=0,\dots,k$, and $K = \{u^{i_0,m_0},\dots,u^{i_n,m_n}\} \in C_\rho$, then

$$\sigma_K = [(u^{i_0},\bar{e}^{m_0}),\dots,(u^{i_n},\bar{e}^{m_n})] \in T_\rho .$$

In this case (cf.[9,11]) $S(C)$ coincides with the space of (discontinuous) piecewise polynomials of (total) degree k with respect to the triangulation Θ. As for the general case we will assume that the n-simplex $\bar{\sigma} = [(u^{i_0,m_0},\bar{e}^{m_0}),\dots,(u^{i_n,m_n},\bar{e}^{m_n})]$ is a 'distortion of $\sigma = [(u^{i_0},\bar{e}^{m_0}),\dots,(u^{i_n},\bar{e}^{m_n})]$ such that

$$\bar{T} = \{ \bar{\sigma} : \sigma \in T_\Theta \}$$

is still a triangulation of $\Omega \times \Sigma^k$. In terms of the knots the points $u^{i,m}$ may be considered as being obtained by 'pulling' the (k+1) fold knot $u^i = (u^i, \bar{e}^m)_{|R^s}$, m=0,...,k, apart. In view of (2.6) the discontinuous piecewise polynomials may be 'smoothed' in this way even into (k-1) times continuously differentiable splines.

Before stating the main results from [11] we have to introduce the following restrictions on the positions of the points $u^{i,m}$ with respect to $u^{i,o}$. Setting for $\rho = [u^{i_o}, \dots, u^{i_s}] \in \Theta$

$$B_\rho = \bigcap \{ [u^{i_o,m}, \dots, u^{i_s,m}] : m=0,\dots,k\}$$

we require

$$(4.14) \quad vol_s(\bigcup \{[u^{i_o,m}, \dots, u^{i_s,m}] : m=0,\dots,k\})/vol_s(B_\rho) < b$$

for some fixed constant $b < \infty$. Hence for any $\rho \in \Theta$ $B_\rho \subset \rho$ is a region of comparable extent where only those B-splines are incident which correspond to knot sets in C_ρ, i.e

$$(4.15) \quad B_\rho \subset \bigcap \{[K] : K \in C_\rho\} \quad , \quad B_\rho \cap [K] = \emptyset \text{ when } K \notin C_\rho .$$

Assuming also that the triangulation Θ of Ω is 'regular' in the sense that the ratios of the circumscribed and inscribed balls of any $\rho \in \Theta$ are uniformly bounded by some constant c we have

<u>THEOREM 4.1</u> [11] <u>Suppose Θ is regular and C satisfies (4.14) then</u>

i) <u>the collection $\{M(x|K) : K \in C\}$ forms a basis for $S(C)$ which is stable in the following sense:</u>

$$d(s,k,b,c,p) \|\{a_K\}\|_{\ell_p} \le \| \sum_{K \in C} a_K M_{K,p}(x) \|_p(\Omega) \le \|\{a_K\}\|_{\ell_p} ,$$

<u>where</u> $M_{K,p}(x) = k!vol_s([K])^{1-1/p} M(x|K)$ <u>and</u> $\|\{a_K\}\|^p_{\ell_p} = \sum_{K \in C} |a_K|^p$.

ii) <u>There are linear projectors</u> $Q(C,f)(x) = \sum_{K \in C} \lambda_K(f) M(x|K)$, $Q(C,\cdot) : L_p(\Omega) \to S(C)$ <u>such that for</u> $\rho \in \Theta$

$$\| f - Q(C,f)\|_p(\rho) \le \gamma \inf\{\| f-g\|_p(\bigcup\{[K] : [K] \cap \rho \neq \emptyset\}) : g \in \Pi_{s,k}$$

where the constant γ depends on s, k, b, c but not on f and $\rho \in \Theta$.

iii) Whenever f and $M(x|K)$, $K \in C$ are sufficiently smooth the following global estimates hold for $|\alpha| \leq k$.

$$\|D^{\alpha}(f - Q(C,f))\|_{p}(\Omega) \leq \kappa\, h^{k+1-|\alpha|}\Big(\sum_{|\beta|=k+1}\|D^{\beta}f\|_{p}^{p}(\Omega)\Big)^{1/p},$$

where $h = \max\{\operatorname{diam}[K] : K \in C\}$ and κ does not depend on f (but on C when $|\alpha| > 0$).

REMARKS i) Thm. 4.1 i) generalizes the well known univariate 'stability estimate' due to de Boor (cf. [2]) since the univariate 'consecutive' B-spline basis is covered by the above construction as a special case. For $s > 1$ a condition like (4.14) turns out to be necessary for the stability of the corresponding B-spline basis [11] . However, it follows from some purely combinatorial criteria [12] that the condition (4.14) is not necessary for the mere linear independence of the B-splines induced by the above construction principle. Another typical result in this context is for instance [12]

$$\dim \operatorname{span}\{M(x|K') : K' \subset K,\ |K'| = m,\ |K| = n\} = \binom{s+n-m}{s}.$$

ii) Note that the restrictions on pulling the knots apart do not interfere in principle with (even optimal) global smoothness requirements nor with strongly varying knot densities. Thus the local estimates in Thm. 4.1 ii) are of particular importance with regard to functions with singularities. Thm 4.1 ii), iii) confirm the optimal local and global approximation properties of linear combinations of multivariate B-splines.

iii) There are various possible choices for the biorthogonal functionals λ_K mentioned in Thm. 4.1 ii), iii). Those proposed in [9,11] involved derivatives and were only used for theoretical purposes. From a practical point of view one may prefer to deal only with point evaluations. To this end, let $Y_{\rho} = \{ y_{\rho}^{j} : j=1,\dots,\binom{s+k}{s} = m\}$ be a k-unisolvend set of points in B_{ρ} (4.14), i.e. there are polynomials $L_i \in \Pi_{s,k}$, $i=1,\dots,m$, such that

$$L_i(y_\rho^j) = \delta_{ij}, \quad i, j = 1,\dots,m = \binom{s+k}{s}.$$

On the other hand, one can find on account of (4.15) and Thm. 4.1 ii) a unique set of coefficients a_{ij}^ρ such that

$$\sum_{j=1}^{m} a_{ij}^\rho \, M(x|K_{j\rho}) = L_i(x), \qquad i = 1,\dots,m, \; x \in B_\rho,$$

for some fixed ordering $\{K_{j\rho}: j=1,\dots,m\}$ of the knot sets in C_ρ. Thus choosing $x = y_\rho^j$, $j=1,\dots,m$, yields

$$(4.16) \qquad A_\rho M_\rho = \mathrm{id}_m = M_\rho A_\rho$$

where $A_\rho = (a_{ij}^\rho)_{i,j=1}^{m}$, $M_\rho = (M(y_\rho^j|K_{i\rho}))_{i,j=1}^{m}$ and id_m is the mxm unit matrix. In particular, we conclude then

$$\sum_{i=1}^{m} a_{ij}^\rho \, M(y_\rho^i|K_{r\rho}) = \delta_{rj}, \qquad r, j = 1,\dots,m,$$

so that

$$\tilde{\lambda}_{K_{j\rho}}(f) = \sum_{i=1}^{m} a_{ij}^\rho \, f(y_\rho^i), \quad j=1,\dots,m,$$

satisfy $\tilde{\lambda}_{K_{j\rho}}(M(\cdot|K_{i\rho})) = \delta_{ij}$, $i,j=1,\dots,m$. Of course, (4.15) affirms then

$$\tilde{\lambda}_K(M(\cdot|K')) = \delta_{K,K'}, \quad K, K' \in C.$$

Clearly, (4.16) says that the interpolation problem with respect to the set of nodes $\bigcup\{Y_\rho : \rho \in \Theta\}$ is well posed. Moreover, the corresponding collocation matrix has a very simple form, namely it is just a block diagonal matrix with blocks M_ρ of the fixed size mxm. So, its inverse is obtained by stacking the corresponding inverse blocks A_ρ and the interpolation reduces to a completely local proceedure whose solution takes $O(|C|)$ operations.

iv) The practical realization of such a setting would essentially consist of the following two steps:

- I. The generation of a suitable triangulation Θ of Ω
- II. pulling the knots apart, i.e. associating with each

vertex of the s-simplices in Θ k further points $u^{i,q}$, q=1,...,k, and forming then the knot sets according to (4.3) or (4.6).

I) may be accomplished (at least for s=2) by one of the various known (interactive or even automatic) proceedures proposed in the recent literature (see e.g. [4, 22]).

II): Interactive graphics terminals seem to be the most suitable hardware devices to deal with pulling the knots apart. In fact, placing the additional points $u^{i,q}$, q=1,...,k, on a terminal screen, say, would allow to control the knot densities and the (local) smoothness properties of the approximating splines directly (at least for two-dimensional problems). However, there seems to be no simple general rule for placing the $u^{i,q}$ automatically so that optimal smoothness is guaranteed. One may of course add to each vertex u^i just k random vectors whose lengths are bounded by a fixed fraction of the minimum of the distances of u^i from the vertices of the adjacent s-simplices.

iv) The choice of the $u^{i,q}$ (and also the structure of Θ) will also affect the number of B-splines incident at a given point x of the domain Ω and thereby the computational work encountered when evaluating a linear combination of B-splines (at least when $x \notin B_\rho$, $\rho \in \Theta$).

v) One has to find suitable numberings for the knot sets in C (in order to minimize for instance the bandwidth of the associated Gram matrix). In view of the known strategies for triangulations (cf. [23]) one may determine first an ordering for the configurations C_ρ, $\rho \in \Theta$, and combine this with a fixed numbering for the knot sets in each C_ρ.

Concerning iv) and v) it is useful to consider a certain subclass of triangulations Θ. To this end, we call an s-polytope C (cf. [14]) an s-quasicube if C is combinatorially equivalent (cf. [14 , p. 38]) to $[0,1]^s$. Such an equivalence is conveniently establihed in terms of a one-to-one correspondence E between all the faces of $[0,1]^s$ and those of C. It is not hard to verify that $K_s(C,E) = \{[E(v^0,\dots,E(v^s)]: [v^0,\dots,v^s] \in K_s\}$ is a triangulation of C, again called Kuhn's triangulation (cf. [11]). For any decomposition $\mathcal{Q} = \{C_i : i=1,\dots,N\}$ of Ω into s-quasi-cubes one can choose $E_i: [0,1]^s \to C_i$ such that

$$(4.17) \qquad \Theta = \bigcup\{K_s(C_i,E_i) : i=1,\dots,N\}$$

is a triangulation of Ω. Analogously we define $K_{s,k}(C_i,E_i)$ as a triangulation of $C_i \times \Sigma^k$ induced by $K_{s,k}$ (4.9). In this case we have

$$T_\Theta = \bigcup\{K_{s,k}(C_i,E_i) : i=1,\dots,N\} .$$

Note that the number of simplices in Θ adjacent at any interior vertex is constant (cf. Fig. 4.1 below).

Concerning the questions iv), v) it is sufficient to consider certain uniform configurations of the above type (4.17), although from a practical point of view this will of course by no means exploit the full flexibility of the concept.

Proceeding similarly as in [8] we let for $h = 1/m$, $m \in \mathbb{N}$

$$\mathcal{Q} = \{ [0,h]^s + h\nu : 0 \le \nu \le \underline{m}=(m,\dots,m) \in \mathbb{Z}^s_+ \}$$

be a partition of $[0,1]^s$ into cubes of side length h. Choosing an sxn matrix of the form

$$(4.18) \qquad H = \begin{pmatrix} 1 & & 0 & c_{11} & \cdots & c_{1k} \\ & \ddots & & \vdots & & \vdots \\ 0 & & 1 & c_{s1} & \cdots & c_{sk} \end{pmatrix}$$

we may associate with each $\sigma_{n,j} \in K_{s,k}$ (4.9), (4.10) the knot set

$$(4.19) \qquad K_j = \{Hv^0_{\pi_{n,j}},\dots,Hv^n_{\pi_{n,j}}\}, \; j=1,\dots,n!/k! .$$

As mentioned before, this may be easily realized by combining (4.9) and a simple algorithm for K_n proposed in [1]. As a special case of (4.12) we obtain the global configuration

$$C_{H,h} = \{K_{h,j,\nu} = h(K_j + \nu) : 0 \le \nu \le \underline{m}\}$$

by scaling and shifting the knot sets K_j. So the structure of

$$S_{H,h} = S(C_{H,h})$$

is completely determined by the matrix H (4.18). Of course, when $c_{ij} = 0$ in (4.18) all the knots would coalesce with the vertices of the cubes $[0,h]^s + h\nu$ and $S_{H,h}$ reduces to the space of (discontinuous) piecewise polynomials with respect to a triangulation of the following type

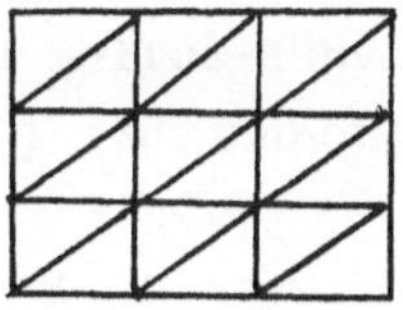

Fig. 4.1

When $c_{ij} \neq 0$ the knots are uniformly pulled apart by H. Specifically, every knot set in $C_{H,h}$ will be in general position iff the vectors

(4.20) $\quad H(v_j^{i_1} - v_j^{i_0}), \ldots, H(v_j^{i_s} - v_j^{i_0})$

are linearly independent for any set of indices $\{i_0, \ldots, i_s\} \subset \{0, \ldots, n\}$ and any $\sigma_{n,j} = [v_j^0, \ldots, v_j^n] \in K_{s,k}$ (cf. (4.9)).

Let

$$J = \{ \{I_i : i=1,\ldots,s\} : I_i \subset \{1,\ldots,n\},\ I_i \cap I_j = \emptyset,\ i \neq j,$$

$$q \in I_{i+1} \cap \{s+1,\ldots,n\} \text{ implies } q > \max\{j \in I_i\} \}$$

and let H^j denote the j-th column of H. Recalling (4.2) and (4.10) it is not hard to realize that (4.20) holds iff the vectors

(4.21) $\left\{ \sum_{i=1}^{s} \sum_{q \in I_i} H^q : \{I_i\} \in J \right\}$ are linearly independent

(cf. [8]). Thus (2.6) yields

<u>PROPOSITION 4.1</u> $\quad S_{H,h} \subset C^{k-1}(R^s)$ <u>iff</u> H <u>satisfies</u> (4.21).

Furthermore, the condition (4.14) may be restated in terms of H by requiring e.g.

$$(4.22)\qquad \sum_{j=1}^{k} |c_{ij}| \leq d < 1/3\,, \qquad i=1,\ldots,s.$$

Concerning iv) the c_{ij} should be all negative so that the knots are pulled towards the lower left quadrant when s = 2. The evaluation of any $S \in S_{H,h}$ at $x \in [0,h]^s + h\nu$ will then involve only the B-splines $M(x|K_{h,j,\mu})$, $\nu \leq \mu$, $\|\mu-\nu\|_\infty \leq 1$, j=1,.. .,n!/k!. Note that in the univariate case $c_{1j} = c_j < 0$ would induce the standard consecutive B-spline basis (cf. [9,11]). In geometrical terms, $c_{ij} < 0$ means that $[0,1]^s \times \Sigma^k$ is tilted by the nxn matrix

$$\bar{H} = \begin{pmatrix} H \\ (\delta_{ij})_{i=s+1,j=1}^{n\quad ,\; n} \end{pmatrix}$$

in such a way that certain 'diagonals' are shortened (see the dashed lines in the following figure for s = 2, k = 1).

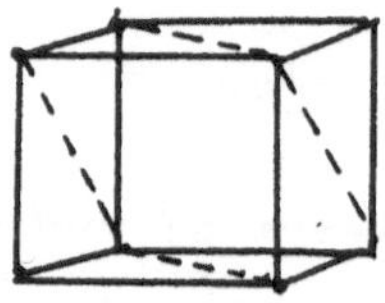

The simplest case s = 1, k = 1 indicates already that then the number of B-splines which are incident at a given point x is minimized

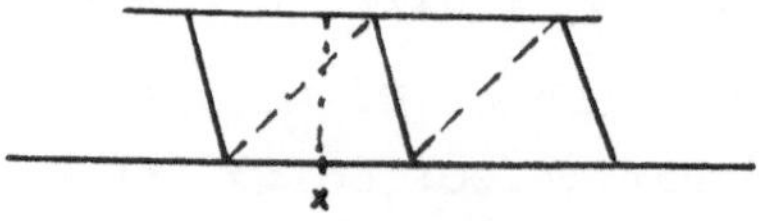

as opposed to

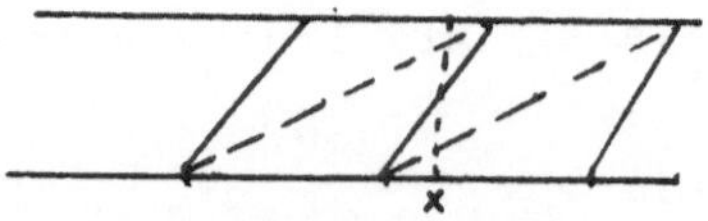

which would correspond to positive c_{ij}. This suggests of course also a preferable choice of the knots $u^{i,q}$, $q > 0$, for non-uniform triangulations Θ of the type (4.17).

A 'natural' ordering for the elements of $C_{H,h}$ would be to number the cubes $[0,h]^s + h\nu$ lexicographically and furthermore, to order the knot sets associated with each cube according to (4.9), (4.19), i.e. when $h = 1/m$

$$(4.23) \qquad \bar{K}_r = K_{h,j,\nu} \quad , \quad r = (n!/k!)(\nu_1+\nu_2 m+\ldots+\nu_s m^{s-1})+j \ .$$

However, when the constant d in in (4.22) is very small so that $S_{H,h}$ is close to the space of discontinuous piecewise polynomials one should prefer an ordering of the following type

$$(4.24) \quad \bar{K}_{r(i-1)+j} = K_{m_j} \ , \ j=1,\ldots,r, \text{ when } T_{\sigma_{s,i}} = \{\sigma_{n,m_1},\ldots,\sigma_{n,m_r}\}, \ i=1,\ldots,s!$$

where $r = \binom{s+k}{s}$, i.e those knot sets are kept together which are associated with the same s-simplex in Θ.

5. NUMERICAL EXAMPLES

Let us discuss now some numerical tests for $s = 2$. In order to establish the Gram matrix G for the normal equations

$$\sum_{j=1}^{n!m^2/2} a_j \, I(\bar{K}_i|\bar{K}_j) \quad = \quad \int_{R^2} f(x)M(x|\bar{K}_i)dx$$

for a fixed numbering $\bar{K}_j$, $j=1,\ldots,n!m^2/2$, of the knot sets, we only have to compute the following blocks of inner products for the knot sets K_j (4.19), i.e. for $h = 1$:

$I(K_i|K_j)$, $i=1,\ldots,n!/2$, $j=i,\ldots,n!/2$;

$I(K_i|K_j+ (1,0))$, $i,j=1,\ldots,n!/2$;

$I(K_i|K_j+ (0,1))$, $i,j=1,\ldots,n!/2$;

$I(K_i|K_j+ (1,1))$, $i,j=1,\ldots,n!/2$;

$I(K_i+ (1.0)|K_j+ (0,1))$, $i,j=1,\ldots,n!/2$.

The corresponding quantities for any other mesh size $h = 1/m$ are in view of (3.6) simply obtained by rescaling. Since the configuration $C_{H,h}$ is induced by a triangulation of $[0,1]^s \times \Sigma^k$ we have because of (4.22)

$$[\bar{K}_j] \cap [\bar{K}_i] \neq \emptyset \quad \text{implies} \quad \bar{K}_j \cap \bar{K}_i \neq \emptyset.$$

Moreover, concerning the first of the above blocks we conclude from (4.1) that $|K_i \cap K_j| \geq 2$, $i,j = 1,\ldots,n!/2$. The remaining blocks are typically very sparse and the intersections of the corresponding B-spline supports use to be again in view of (4.22) relatively small. Recalling the reasoning at the end of the previous section this suggests to apply Algorithm C in order to compute the above inner products.

As a case of particular interest we consider $s = 2 = k$. In fact, choosing for instance

$$H = \begin{pmatrix} 1 & 0 & -0.1 & -0.3 \\ 0 & 1 & -0.17 & -0.08 \end{pmatrix}$$

the elements of the space $S_{H,h}$ are C^1-splines of degree two (cf Prop. 4.1) which may therefore also give rise to a conforming finite element scheme for fourth order problems.

We have used (3.5) together with a third degree and a fifth degree cubature formula with positive coefficients (see Stroud [24, p. 308, formula 3-4 , p. 312, formula 5-1]) for the computation of the right hand side coefficients of the above normal equations. As expected, a third degree formula is certainly sufficient for $k \leq 3$ and smooth functions f. Indeed for the above choice of H and even $h = 1/3$ the respective results differe typically by less than 0.01.

For $\dim S_{H,h} = 108$, H as above, the ratios of the maximal and minimal eigenvalues of the Gram matrix G are bounded by 17 when using an ordering of the type (4.24), namely

$\bar{K}_1 = K_7$, $\bar{K}_2 = K_8$, $\bar{K}_3 = K_9$, $\bar{K}_4 = K_{10}$, $\bar{K}_5 = K_{11}$, $\bar{K}_6 = K_1$, $\bar{K}_7 = K_2$, $\bar{K}_8 = K_3$ $\bar{K}_9 = K_4$, $\bar{K}_{10} = K_5$, $\bar{K}_{11} = K_6$, $\bar{K}_{12} = K_{12}$.

The condition number went up to about 22 when using the ordering (4.23). The normal equations were solved by band Cholesky algorithms for

$$f_1(x) = \cos(7(x_1+x_2)) + (0.1+(x_1-0.5)^2+(x_2-0.5)^2)^{-1}$$

and

$$f_2(x) = \cos(7(x_1+x_2)) + |x_1 - x_2|^{1/2}.$$

Whereas f_1 is very smooth, f_2 has singular first order derivatives along the main diagonal of the unit-square.

The maximal pointwise errors produced by the least squares approximants with the above H are for $h = 1/10$

$$E_1 \doteq 0.0178 \quad \text{and} \quad E_2 = 0.0988,$$

respectively. The behavior of the approximations is illustrated by the following figures. These plots were obtained by evaluating the functions and their spline fits at the lattice points of uniform rectangular grids. The evaluation of the splines is of course based on the recurrence relation (2.3). Since the configurations C, and in particular $C_{H,h}$, are induced by triangulations different knot sets will frequently have common knots.
Hence a considerable amount of computational work can be saved by exploiting the fact that the recursion for different B-splines may involve common lower order B-splines. This suggests to use special recursion codes e.g. for the linear combinations of all those B-splines associated with the same (quasi) cube or, more generally, with the same configuration C_ρ, $\rho \in \Theta$. Note that such algorithms will be also completely determined by the combinatorial structure of the associated 'local' configurations C_ρ. Hence they will be applicable to more general non-uniform knot configurations as well.

As mentioned above the following figures show continuously differential least squares fits to the functions f_1, f_2 for $h = 0.1$ and H as above (noting that f_2 has singular derivatives along the diagonal of the unit square).

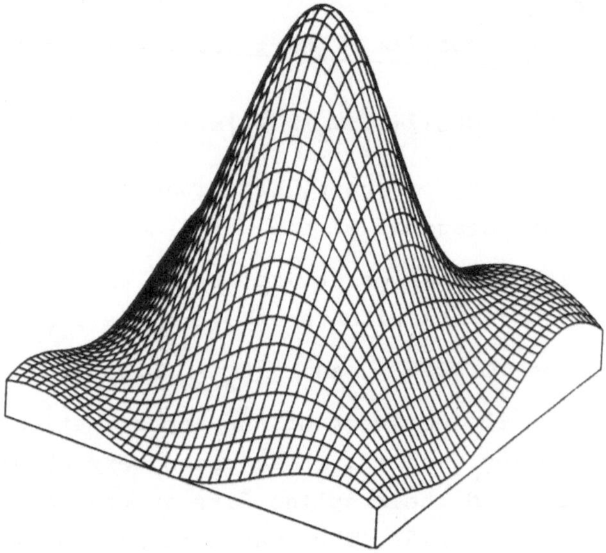

Fig.5.1

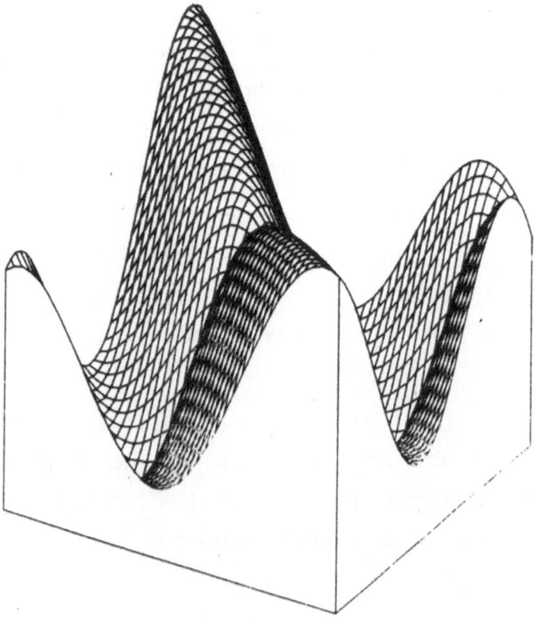

Fig.5.2

REFERENCES

[1] Allgower, E., Georg, K.: Triangulations by reflections with applications to approximations. In: Numerische Methoden der Approximationstheorie Bd. 4, ed. L. Collatz, G. Meinardus, H. Werner, Birkhäuser Basel, 1978, 10-32.

[2] de Boor, C.: Splines as linear combinations of B-splines. In: Approximation Theory II, ed. G.G. Lorentz, C. K. Chui, L. L. Schumaker, Academic Press, 1976, 1-47.

[3] de Boor, C., Lyche, T.,Schumaker, L. L.: On calculating with B-splines II. Integration. In: Numerische Methoden der Approximationstheorie, ed. L. Collatz, G. Meinardus, H. Werner, Basel, Birkhäuser, 1976, 123-146.

[4] Cavendish, J.C.: Automatic triangulation of arbitrary planar domains for the Finite Element Method. International Journal for Numerical Methods in Engeneering, 8 (1974), 679-696.

[5] Dahmen, W.: On multivariate B-splines. SIAM J. Numer. Anal. 17(1980), 179-191.

[6] Dahmen, W.: Multivariate B-splines - recurrence relations and linear combinations of truncated powers. In: Multivariate Approximation Theory, ed. W. Schempp, K. Zeller, Basel, Birkhäuser, 1979, 64-82.

[7] Dahmen, W.: Konstruktion mehrdimensionaler B-Splines und ihre Anwendungen auf Approximationsprobleme. In: Numerische Methoden der Approximationstheorie, ed. L. Collatz, G. Meinardus, H. Werner, Basel, Birkhäuser, 1980, 84-110.

[8] Dahmen, W.: Approximation by linear combinations of multivariate B-splines. To appear in J. Approx. Theory.

[9] Dahmen, W.: Multivariate B-Splines - ein neuer Ansatz im Rahmen der konstruktiven mehrdimensionalen Approximationstheorie. Habilitationsschrift, Bonn, Wintersemester 80/81.

[10] Dahmen, W., Micchelli, C.A.: Computation of inner products of multivariate B-splines. To appear in Numerical Functional Analysis and Applications.

[11] Dahmen, W., Micchelli, C.A.: On the linear independence of multivariate B-splines I. Triangulations of simploids. To appear.

[12] Dahmen, W., Micchelli, C.A.: On the linear independence of multivariate B-splines II. Complete configurations. In preparation, ms.

[13] Goodman, T.N.T., Lee, S.L.: Spline approximation of Bernstein Schoenberg type in one and two variables. To appear.

[14] Grünbaum, Branko: Convex Polytopes, London, Interscience, 1967.

[15] Hakopian, H.: On multivariate B-splines. To appear.

[16] Höllig, K.: A remark on multivariate B-splines. To appear in J. Approx. Th.

[17] Höllig, K.: Multivariate Splines. To appear.

[18] Johnson, S.M.: Generation of permutations of adjacent transpositions. Math. Comp. 17(1963), 282-285.

[19] Kuhn, K.W.: Some combinatorial lemmas in topology. IBM J. Research and Develop. 45(1960), 518-524.

[20] Micchelli, C.A.: A constructive approach to Kergin interpolation in R^k: multivariate B-splines and Lagrange interpolation. University of Wisconsin Madison, Mathematics Research Center, Technical Summary Report, 1978, to appear in Rocky Mountain Journal of Mathematics.

[21] Micchelli, C.A.: On a numerically efficient method for computing multivariate B-splines. In: Multivariate Approximation Theory, ed. W. Schempp, K. Zeller, Basel, Birkhäuser, 1979, 211-248.

[22] Quesenberry, N.A.: Automatic triangulation of circles and ellipses. Thesis, Fort Collins, 1980.

[23] Schwarz, Hans Rudolf: Methode der Finiten Elemente, Stuttgart, Teubner, 1980.

[24] Stroud, A.H.: Approximate Calculation of Multiple Integrals. Englewood Cliffs, N.Y., Prentice-Hall, 1971.

Wolfgang Dahmen
Institut für Angewandte Mathematik der Universität Bonn
Wegelerstraße 6
53 Bonn, W-Germany

Charles A. Micchelli
IBM Thomas J. Watson Research Center
P.O. Box 218
Yorktown Heights, N.Y. 10598
U.S.A.

EXCHANGE ALGORITHM FOR MULTIVARIATE POLYNOMIALS

Ph. DEFERT - J.-P. THIRAN

Department of Mathematics
Facultés Universitaires de Namur
Rempart de la Vierge 8
B-5000 Namur Belgium

In [10], Töpfer proposed an algorithm for linear Chebyshev approximation without Haar condition, which has been thoroughly investigated in two recent contributions of Carasso and Laurent [2], [3].

The aim of this paper is twofold. First, it presents a modified version of the Carasso-Laurent algorithm for a discrete Chebyshev problem. In particular, when uniqueness does not occur, it computes the strict approximation [9] which, in some sense, is an optimal way of choosing one solution among all best approximations. Secondly, as a natural consequence of the proposed algorithm, one considers the minimal H-sets introduced by Collatz [4], by giving a complete classification of so-called degenerate minimal H-sets for two-variable polynomials of any degree.

1. The algorithm.

1.1. Basic concepts.

Let F be a linear subspace of C(Q), with Q a compact of $\mathbb{R}^p$, spanned by $f_1, f_2, \ldots, f_n$, and g a real continuous function on Q. The best approximation of g in F on Q is the element f_o of F such that

$$\|f_o - g\| = \inf_{f \in F} \|f - g\|$$

where
$$\|h\| = \sup_{p \in Q} |h(p)| \quad .$$

The classical way of computing best approximation is the exchange algorithm which may fail if Haar's condition is not satisfied. To overcome this problem, Carasso and Laurent ([2] , [3]) investigated a generalized algorithm first introduced by Töpfer ([10]) and based on the concept of chain of supports.

A subset $S = \{p_1,\dots,p_{m+1}\}$ of Q is said to be a support ([3]) of V, a subspace of $\mathbb{R}^n$, if there exist real numbers $\lambda(p_i)$, $i = 1,\dots,m+1$, not all zero such that

$$\sum_{i=1}^{m+1} \lambda(p_i)\, u(p_i) \in V$$

where $u(p) = (f_1(p),\dots,f_n(p))^T$. S is minimal when no proper subset of it is a support too, and this implies that every $\lambda(p_i)$ is non zero and $m \leqslant n - \dim V$. If S is a support of the 0-vector space, it is simply a H-set ([4] , [6] , [1]) and the signs of the $\lambda(p_i)$ form the associated sign pattern.

One can combine supports to build a regular chain in the following way. Let $S_1 = \{p_{1,1},\dots,p_{1,m_1+1}\}$ be a minimal H-set relative to F satisfying

$$\sum_{i=1}^{m_1+1} \lambda_1(p_{1,i})\, u(p_{1,i}) = 0 \quad . \tag{1}$$

The set of vectors $a \in \mathbb{R}^n$ which achieve the best approximation of g in F on S_1 is the affine variety of dimension $n - m_1$

$$W_1 = \{a \in \mathbb{R}^n : e(p_{1,i}) = \eta_1 \operatorname{sign} \lambda_1(p_{1,i}).\alpha_1\}$$

where $$e(p) = \sum_{i=1}^{n} a_i\, f_i(p) - g(p) \quad , \quad |\eta_1| = 1 \quad ,$$

and $$\alpha_1 = \min_{a \in \mathbb{R}^n} \max_{p \in S_1} |e(p)| \quad ,$$

called the first support deviation. The linear subspace V_1 of $\mathbb{R}^n$ orthogonal to W_1 is

$$V_1 = \text{span}\,\{u(p_{1,i}) : i = 1, \ldots, m_1 + 1\} \quad .$$

Now let $S_2 = \{p_{2,1}, \ldots, p_{2,m_2+1}\}$ be a minimal support of V_1, i.e. a minimal H-set relative to W_1, satisfying

$$(2) \qquad \sum_{i=1}^{m_2+1} \lambda_2(p_{2,i})\, u(p_{2,i}) \in V_1 \quad .$$

The set of best approximations of g in W_1 on S_2 is

$$W_2 = \{a \in W_1 : e(p_{2,i}) = \eta_2 \text{ sign } \lambda_2(p_{2,i}).\alpha_2\}$$

where $|\eta_2| = 1$ and $\alpha_2 = \min_{a \in \mathbb{R}} \max_{p \in S_2} |e(p)|$

is the second support deviation. The linear subspace orthogonal to W_2 is

$$V_2 = \text{span}\,\{u(p_{1,i}) : i = 1, \ldots, m_1+1 \,;\, u(p_{2,i}) : i = 1, \ldots, m_2+1\} \; .$$

Repeating this process, one generates step by step a chain of supports $C = (S_1, \ldots, S_M)$ with the corresponding deviations vector $(\alpha_1, \ldots, \alpha_M)$ until the variety W_M is reduced to one vector $\bar{a}$ called the solution of the chain. Every support made of a single point can be deleted without changing the involved spaces and the solution, in this way, one obtains a regular chain.

Example. If $F = P_2^1 = \text{span}\,\{1, x, y\}$ in $C(Q)$, $Q \subset \mathbb{R}^2$, three points one the x-axis (more generally on a straight line) form a first support S_1 ([6]). The space W_1 is of the form

$$W_1 = \{a^* + \lambda\, (0, 0, 1)^T : \lambda \in \mathbb{R}\}$$

where a^* is a particular solution, and the set $V_1 = \{(\mu, \nu, 0)^T : \mu, \nu \in \mathbb{R}\}$. Two points outside the x-axis compose the second support.

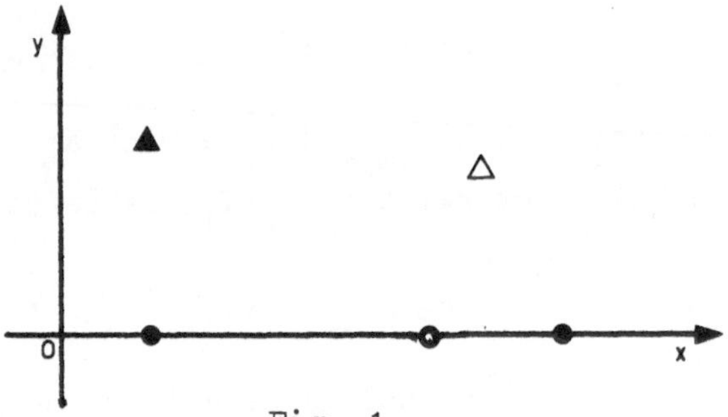

Fig. 1.

1.2. The Carasso and Laurent algorithm.

It is based on the Generalized Exchange Theorem ([3]).

Let $C=(S_1,\dots,S_M)$ be a chain. From two successive supports of C, S_i and S_{i+1} such that $\alpha_i < \alpha_{i+1}$, one can partition S_i in $B_i \cup C_i$ to create a new chain $\tilde{C}=(S_1,\dots,\tilde{S}_i,\tilde{S}_{i+1},S_{i+2},\dots,S_M)$ with $\tilde{S}_i = B_i \cup S_i$ and $S_{i+1} = C_i \neq \emptyset$ in such a way that $\tilde{\alpha}_i > \alpha_i$.

In the foregoing developments, every occurence of the word "exchange" will refer to this theorem.

The algorithm constructs a sequence of regular chain $C^i=(S_1^i,S_2^i,\dots,S_{M_1}^i)$. Given C^i, one constructs C^{i+1} in the following manner : One first determines the extremal value of the error function $\overline{e}^i$ corresponding to the solution $\overline{a}^i$ of C^i

$$A^i = |\overline{e}^i(q^i)| = \max_{p\in Q} |\overline{e}^i(q^i)| \quad .$$

Three different cases are to be considered :

1. If $\alpha_1^i = A^i$, the vector $\overline{a}^i$ is a best approximation;
2. If $\alpha_j^i = A^i$, $j \geq 2$, one has to exchande S_{j-1}^i and S_j^i;
3. If $A^i > \alpha_j^i$ for all j, one computes the smallest index j such that $(S_1^i,\dots,S_{j-1}^i,\{q^i\},S_j^i,\dots,S_{M_i}^i)$ is still a chain and one exchanges S_{j-1}^i and $\{q^i\}$.

<u>Example</u>. In P_2^1, let us take $C^i = (S_1, S_2)$ of Figure 1. If $\|\overline{e}\| = \alpha_2 > \alpha_1$, we are in Case 2 and S_1, S_2 are to be exchanged so that C^{i+1} is reduced to one single support of four points (see Figure 2).

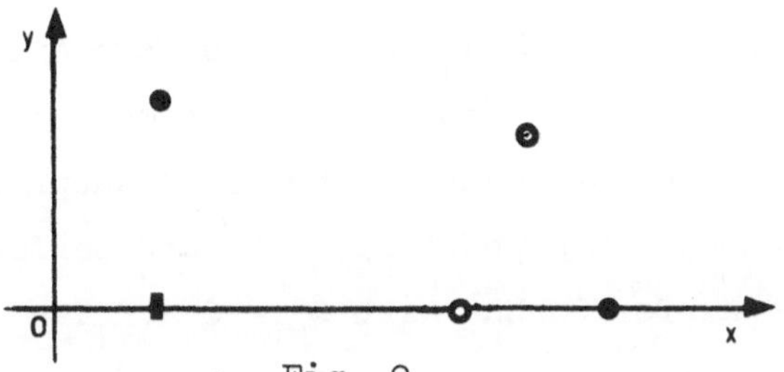

Fig. 2.

Now, if $\alpha_1 < \alpha_2$ and $|\overline{e}(q^i)| = \|\overline{e}\|$, we are in Case 3; when q^i is on the x-axis, $\{q^i\}$ is exchanged with S_1 and in the other case, S_2 is concerned.

The algorithm is constructing a sequence of chains, in a way that the deviations vectors are lexicographically increasing and this implies the convergence of the algorithm ([3]). But in this procedure, the highest lexicographic increase of the deviations vector is obtained when the exchange is concerning the support of the lowest possible index; in our example, in Case 3, one should first examine the points of the x-axis, in other words, the points p such that $u(p) \in V_1^i$. This motivates the following modification.

1.3. <u>The modified algorithm</u>.

Instead of generating a sequence of regular chains, it constructs one single chain, corresponding to a best approximation, in the following way :

One starts with a minimal H-set S_1.

<u>Step 0</u>. One determines the set P_1 of all $p \in Q = Q_1$ such that $u(p) \in V_1$ ..

Step 1. One computes the deviation α_1, the variety W_1, and finds $q \in P_1$ if possible such that

$$A = |e(q)| > \alpha_1 .$$

Step 2. If $A > \alpha_1$, S_1 and $\{q\}$ are exchanged to get a new $\tilde{S}_1$, to which is associated the same P_1; then one goes to Step 1. In the other case, the first subproblem is solved and one chooses a support S_2 of V_1 and performs the same computations as for S_1 but with $Q_2 = Q_1 \setminus P_1$.

In this process, if α_2 becomes larger than α_1, one exchanges S_1 and S_2 to get new $\tilde{S}_1$ and $\tilde{S}_2$ and one goes to Step 0.

One finally obtains a regular chain $C = (S_1, \ldots, S_M)$ with $\alpha_1 \geqslant \alpha_2 \geqslant \ldots \geqslant \alpha_M$ and

$$\alpha_i = \max_{p \in Q_i} |e(p)| \quad ; \quad Q_i = Q \setminus \bigcup_{j=1}^{i-1} P_j .$$

If Q is a finite points set, this construction follows presicely the definitions of strict approximation ([9] , p. 240) which, in case of non-uniqueness, singles out one optimal solution as the unique best of the best approximations.

The main problem introduced by the modified algorithm consists in finding the subsets P_i of Q_i. It can be settled as follows. Let $S = \{p_1, p_2, \ldots, p_{m+1}\}$ be a minimal H-set relative to F; the linear system

$$\sum_{i=1}^{m+1} \lambda(p_i)\, f_j(p_i) = 0 \qquad (j = 1, \ldots, n) ,$$

in the unknowns $\lambda(p_i)$ has a solution space of dimension 1 so that the transposed system in $a \in \mathbb{R}^n$

$$\text{(3)} \qquad \sum_{i=1}^{n} a_i\, f_i(p_j) = 0 \qquad (1 \leqslant j \leqslant m+1) ,$$

has a solution space of dimension n-m. Let $a^{(1)}, \ldots, a^{(n-m)}$

be a basis of the solution space. The subset P of Q is given by the intersection of n - m curves

$$C^{(j)} = \{p \in Q \; ; \; [u(p)]^T a^{(j)} = 0\}, \quad 1 \leq j \leq n-m \; .$$

If F is a two-variable polynomial space, the $C^{(i)}$ are algebraic curves which can be factorized as follows :

$$C^{(i)} = \Gamma^{(i)} \, C$$

with no common part between the $\Gamma^{(i)}$. The set P is then composed of a common curve C and a set of isolated points $I = \bigcap_{i=1}^{n-m} \Gamma^{(i)}$.

Example. It will be settled later that a set S composed of 14 points on the intersection of two-fourth order curves and 6 points on a straight line form a minimal H-set relative to $P_2^5 = \text{span}\,\{x^i \, y^j \; : \; i+j \leq 5\}$. The set P is the union of the whole straight line and the set of 16 intersection points of the two-fourth order curves (see Fig. 3).

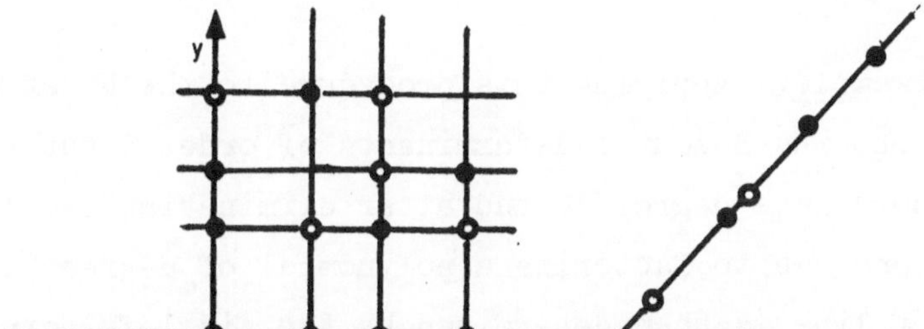

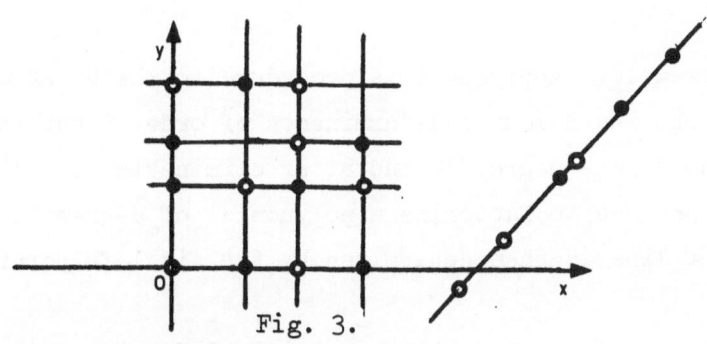

Fig. 3.

1.4. Computational considerations.

To a minimal H-set $S_1 = \{p_{1,1}, \ldots, p_{1,m_1+1}\}$, one can associate a "canonical basis" (Töpfer [10]) $\{u_1, \ldots, u_n\}$ such that

$$u_j(p_{1,i}) = \delta_{ij} \; , \quad j = 1, \ldots, m_1 \; , \quad i = 1, \ldots, m_1 \; ,$$

and $\quad u_j(p_{1,i}) = 0 \; , \quad j = m_1+1, \ldots, n \; , \quad i = 1, \ldots, m_1+1 \; .$

In the implementation of the modified algorithm, it is most

suitable to use these canonical bases. Indeed, they are very easily obtained by a procedure similar to a Gaussian elimination applied to system (3). The first m functions u_i, $i=1,\dots,m$, give immediately the numbers $\lambda_1(p_{1,i})$, $i=1,\dots,m+1$, of relation (1), the quantity α_1 and the variety W_1. Moreover, $u_i(p)=0$, $i=m+1,\dots,n$, are the equations of the n-m curves which determine P_1.

To compute the intersection of two algebraic curves, $\sum_{i+j\leq r} a_{ij}^{(k)} x^i y^j = 0$, $k=1, 2$, the first variable x is eliminated by the construction of the Bezout's determinant. If s is the highest order of the principal minor which is not vanishing, $r-s$ is the degree in x of the common curve C. The roots of this minor yield the isolated points. In fact, one can spare the computation of some roots by eliminating first the known zeros which correspond to the points of S which belong to I. The same operations are performed with the variable y. The common curve is then determined by the points of $S\setminus I$.

Example. Applying this procedure to the H-set of Fig. 3, one obtains Bezout's determinants of order 5 but rank 4; the minors have degree 16 and after eliminating the 14 known roots, one has to factorize a polynomial of degree 2. The straight line is then determined by the six left points.

2. Theoretical considerations on minimal H-sets.

There arises some difficulty in finding P if the number of points of a minimal H-set is less than the maximal value $n+1$: such minimal H-sets, which produce chains composed of several supports, will be called degenerate. This section investigates all possible classes of degenerate minimal H-sets relative to algebraic polynomials of total degree r in two variables x and y

$$(4) \qquad p_r(x,y) = \sum_{i=0}^{r} \sum_{j=0}^{r-i} a_{ij}\, x^i\, y^j$$

for which one has $n = (r+1)(r+2)/2$ and $u(p) = (1\ x\ y\ \dots\ y^r)^T$ with $p = (x,y) \in \mathbb{R}^2$. The curves (3) are then algebraic curves of order r and P is an algebraic variety.

To comply with the terminology used in [5, p. 380], we introduce the following definitions. A system G of points $p_1, \dots, p_\ell$ on a curve C_r is said to be normal for order r and the points are called themselves independent if the vectors $u(p_1), \dots, u(p_\ell)$ are linearly independent. If the maximum number of independent vectors is $\ell - s$ with $s > 0$, the system is superabundant for order r, of superabundance s. When there exists no $p \in \mathbb{R}^2 \setminus G$ such that $u(p) \in \text{span}\,\{u(p_1),\dots,u(p_\ell)\}$, G is complete. For the problem at hand, a degenerate minimal H-set $\{p_1,\dots,p_{m+1}\}$ forms a superabundant system for order r, of superabundance one and the set P is the complete superabundant system characterized by the intersection of $n-m$ independent curves $C_r^{(1)}, \dots, C_r^{(n-m)}$.

If $m+1 = n$, the algebraic variety P amounts to one single curve of order r. If $m+1 < n$, we first assume that the $n-m$ curves have no common part. For $m+2 = n$, due to Bezout's Theorem [5, p. 10], the set P contains at most r^2 different points in the real plane. This number will be taken as r^2 which means, in particular, that all intersection points are simple on both curves. As a consequence, the minimal H-set consists of $n-1$ intersection points and the superabundance of P is $\sigma = (r-1)(r-2)/2$. For $m+2 = n-t$ with $t > 0$, in the $t+2$ curves, it is always possible to select two curves, for instance $C_r^{(1)}$ and $C_r^{(2)}$, without common part. Among their intersection points $p_1, \dots, p_{r^2}$, we shall determine a complete superabundant subset, having superabundance s, $A = \{p_1,\dots,p_{\rho-t};$

$p_{\rho-t+1},\dots,p_{\rho-t+s}\}$ in which $\{p_1,\dots,p_{\rho-t}\}$ forms a normal system and $\rho = n-2$. By [5, p. 382], [7], the superabundance s is the number of independent curves of order $r-3$ which pass through the remaining points $B = \{p_i \,;\, \rho-t+s < i \leqslant r^2\}$: the system A will have superabundance s for order r if the complementary system B has superabundance t for order $r-3$. Therefore, one builds all possible superabundant systems for order r, by taking a minimal H-set relative to degree $r-3$

$$B = \{q_1, q_2, \dots, q_{h+1}\} \quad , \quad h+1 \leqslant \sigma \quad , \quad t=1 \quad , \tag{5}$$

which defines $s = \sigma - h$ independent curves of order $r-3$. This H-set is complementary to a complete superabundant system $A = \{p_1,\dots,p_{r^2-h-1}\}$ of superabundance s for order r, which defines $n-3$ independent curves of order r. If Q denotes the complete superabundant system for order $r-3$, generated by the minimal H-set (5), by adding successively $q_{h+2}, \dots \in Q$ to B, one gets complementary sets of complete superabundant sets for order r. These sets have identical superabundance s but different number of points r^2-h-2, r^2-h-3, ..., thereby producing $n-4$, $n-5$, ... independent curves of order r. If the points of B lie on a curve of order $k \leqslant r-3$, their number must be less than rk since, for card $B > rk$, due to Bezout's theorem, C_k is common to $C_r^{(1)}$ and $C_r^{(2)}$.

Finally, if the $n-m$ curves defining P have a common part C_{r-k} they can be written as $C_{r-k}\, C_k^{(i)}$ $(1 \leqslant i \leqslant n-m)$. If $F = \bigcap_{i=1}^{n-m} C_k^{(i)} = \emptyset$, it is easily verified that the minimal H-set consists of $n-\ell+1$ points on C_{r-k} with $\ell = (k+1)(k+2)/2$. If $F \neq \emptyset$, the minimal H-set contains some points $p_1, \dots, p_g$ on C_{r-k} and some points $q_1, \dots, q_h$ on F, such that one has $g \leqslant n-\ell$ and $h \leqslant H$ where H is the maximum number of independent points on F. It can be shown that there exist different minimal H-sets generating $C_{r-k} \cup F$, which are obtained by choosing

$n-\ell+h$ points from the set $\{p_1,\ldots,p_{n-\ell},q_1,\ldots,q_H\}$ in all possible ways.

To illustrate the foregoing analysis, all classes of degenerate minimal H-sets up to degree four are listed in Table 1 in which H_r^j stands for the j-th H-set relative to degree r and P_r^j is its corresponding variety. In general, the $m+1$ points of H_r^j are divided into c points on a curve C and i isolated points belonging to the intersection I of several curves without common part. Therefore, at Step 2 of the modified algorithm, it will be necessary to examine the whole curve C and $N = \text{card } I - i$ points of the set I. As an example, on the leaf of Descartes $xy = x^3 + y^3$ whose parametrical expression is $x = t/(1+t^3)$ and $y = t^2/(1+t^3)$, a polynomial (4) of third degree becomes

$$p_3(t/(1+t^3)\,,\,t^2/(1+t^3)) = \sum_{i=0}^{9} a_i\ t^i/(1+t^3)^3\ ;\ a_0 = a_9\ .$$

To construct a minimal H-set indexed as H_3^1 in Table 1, we take $t_1 < t_2 < \ldots < t_{10}$ to get

$$\sum_{k=1}^{10} \lambda_k\ v(t_k) = 0\ ;\ v(t) = (1+t^3)^{-3}\ (1+t^9\ t\ t^2 \ldots t^8)^T\ ,$$

where the characteristic coefficients are readily computed as

$$\lambda_k = (-1)^k \prod_{i\neq k} (1+t_i)^{-3} \prod_{i<j,\,i\neq k} (t_i - t_j)\Big(\prod_{i\neq k} t_i + 1\Big)\ .$$

If $\prod_{i\neq k} t_i \neq -1$ for all k, the ten points have alternating signs except for those enclosing the point at infinity $(t=-1)$, the double point at the origin $(t=0)$ and the point given by $t=T$, where no sign changes occur. In case of $\prod_{i\neq \ell} t_i = -1$ for some ℓ, the H-sets amounts to nine isolated points t_k $(1 \leq k \neq \ell \leq 10)$ which form the complete intersection of the leaf of Descartes with another independent curve of order three : it thus belongs to the class H_3^4 of Table 1.

r	j	m + 1	C	c	I	i	N
1	1	3	C_1	3	-	-	-
2	1	6	C_2	6	-	-	-
	2	4	C_1	4	-	-	-
3	1	10	C_3	10	-	-	-
	2	8	C_2	8	-	-	-
	3	5	C_1	5	-	-	-
	4	9	-	-	$\bigcap_{i=1}^{2} C_3^{(i)}$	9	0
4	1	15	C_4	15	-	-	-
	2	13	C_3	13	-	-	-
	3	10	C_2	10	-	-	-
	4	6	C_1	6	-	-	-
	5	14	C_1	5	H_3^4	9	0
	6	14	-	-	$\bigcap_{i=1}^{2} C_4^{(i)}$	14	2
	7	13	-	-	$\bigcap_{i=1}^{3} C_4^{(i)}$	13	0
	8	12	-	-	$\bigcap_{i=1}^{4} C_4^{(i)}$	12	0

Table 1.

3. Numerical results.

The modified algorithm has been applied to several examples published by Watson in [11]. He used a Remez type algorithm based on linear programming which computes an error function which attains its extremum in exactly $n+1$ points. Therefore, some difficulties arise in singular problems though our modified algorithm is ending without any problem.

Example 1.

$$g(x, y) = \exp(-x^2 - y),$$

$$Q = [0, 1] \times [0, 1], \quad F = P_2^2 = \text{span}\{1, x, y, x^2, xy, y^2\}.$$

The algorithm is ending with a chain of two supports as shown below :

$\alpha_1 = .027275$		$\alpha_2 = .027274$	
S_1		S_2	
x	y	x	y
.272	0	1	.622
.836	0	0	.216
1	.62		
.676	1		
0	1		
0	.2176		

Table 2.

Example 2.

$$g(x\,,y) \;=\; \sin\,(x^2 + y) \quad ,$$

$$Q = [\,-1\,,\,1\,]\times[\,-1\,,\,1\,] \quad , \quad F = \text{span}\,\{x^i\,y^j \;:\; 0 \leqslant i,j \leqslant 2\}\;.$$

Our algorithm gives the following chain of two supports :

$\alpha_1 = .071228$		$\alpha_2 = .071228$	
S_1		S_2	
x	y	x	x
1	1	-1	1
1	-.13	-1	-.13
1	-1	-1	1
0	1		
0	-.324		
0	-1		
.686	.7		
-.686	.7		

Table 3.

One remarks a symmetry in the problem with respect to the y-axis so that the problem can be simplified in :

$$Q = [\,0\,,\,1\,]\times[\,-1\,,\,1\,] \quad , \quad F = \text{span}\,\{1,y,x^2,x^2y,y^2,x^2y^2\}$$

and the resulting chain is composed of the single support S_1.

References.

[1] M. Brannigan, H-Sets in Linear Approximation. J. Approx. Th. 20 (1977), 153-161.

[2] C. Carasso and P.J. Laurent, Un Algorithme de Minimisation en Chaîne en Optimisation Convexe. SIAM J. Control Optimization 16 (1978), 209-235.

[3] C. Carasso and P.J. Laurent, Un Algorithme Général pour l'Approximation au Sens de Tchebycheff de Fonctions Bornées sur un Ensemble Quelconque. Lecture Notes in Mathematics 556 (1976), Springer Verlag, Berlin.

[4] L. Collatz, Inclusion Theorems for the Minimal Distance in Rational Tschebyscheff Approximation with Several Variables. Approximation of Functions, Garabedian, ed., Elsevier, Amsterdam, 1965, 43-56.

[5] J.L. Coolidge, A Treatise on Algebraic Plane Curves. Dover, New York, 1959.

[6] C. Dierieck, Some Remarks on H-Sets in Linear Approximation Theory. J. Approx. Th. 21 (1977), 188-204.

[7] B. Gambier, Système Linéaire de Courbes Algébriques de Degré Donné. Annales Ecole Normale, Série 3, T. 41 (1924), 147-264.

[8] S.Y. Ku, R.J. Adler, Computing Polynomials Resultants : Bezout's Determinant vs. Collins' Reduced p.r.s. Algorithm. Comm. ACM 12 (1969), 23-30.

[9] J.R. Rice, The Approximation of Functions, Vol. 2 : Nonlinear and Multivariate Theory. Addison-Wesley, Reading, Mass., 1969.

[10] H.J. Töpfer, Tschebyscheff-Approximation und Austauschverfahren bei Nicht Erfüllter Haarscher Bedingung. Tagung, Obserwolfach (1965), ISNM7, Birkhäuser Verlag, Basel, 1967, 71-89.

[11] G.A. Watson, A Multiple Exchange Algorithm for Multivariate Chebyshev Approximation. SIAM J. Numer. Anal. 12 (1975), 46-52.

CONDITIONS FOR A SMOOTH JUNCTION BETWEEN THREE QUASI-RECTANGULAR PATCHES

David Handscomb

In the context of the representation of C^2 three-dimensional surfaces by vector-valued biparametric piecewise polynomials, necessary conditions are obtained for three such surfaces to meet with C^2 continuity along their common parameter lines, with particular attention to the smoothness at the common vertex. The conditions can not be met by pure bicubic splines, but can be met by increasing the polynomial order to six in the three patches surrounding the common vertex. By way of illustration, a spherical octant is represented by just three patches.

The problem to be described here is one that arises in the field of computer-aided design and manufacture of car bodies; similar problems can arise in any field that involves the numerical representation of "sculptured surfaces" - surfaces that can not conveniently be represented as distortions of a single plane.

If a surface is near to a plane then, taking (x,y,z)-axes in and normal to that plane, we can represent it by a function of two variables, $z = f(x,y)$, and approximate it numerically by a bicubic spline, so that f is approximated by a patchwise bicubic polynomial in x and y (the rectangular patches defined by the boundary lines x = constant and y = constant), with C^2 continuity across the patch boundaries. Some more general surfaces we can represent by a vector-valued function of two parameters, $\underline{r} = (x,y,z) = \underline{f}(u,v)$, and approximate by a vector-valued bicubic spline in u and v - that is to say, a bicubic polynomial with vector-valued coefficients on patches with boundary lines u = constant and v = constant. [The part of such a surface formed by a rectangle in (u,v) space is what we mean by a "quasi-rectangular patch".] Notice that a surface of the latter form may have a singular point if anywhere the partial derivatives $\underline{f}_u$ and $\underline{f}_v$

are parallel or vanish. As long as we guard against this possibility, we may apply this scheme to any surface that we can divide into four-sided patches meeting by fours at common vertices.

It is impossible, however, to divide a closed surface in this fashion, and is often undesirable even if the surface is not closed; we may, for instance, need to have only three patches meeting at some vertex, as shown in Figure 1. (There are other ways in which irregul-

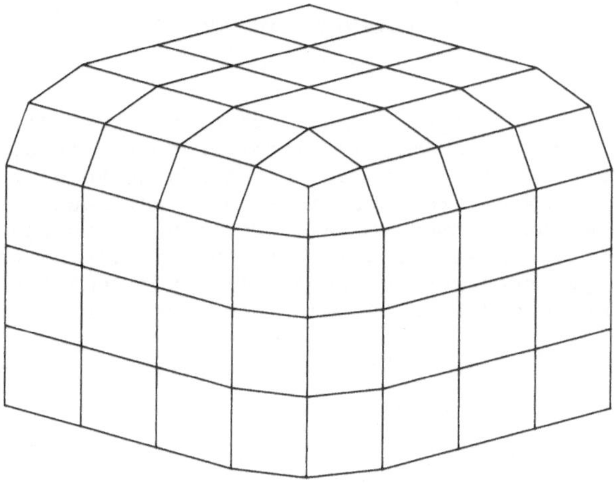

Figure 1. Possible subdivision at a corner

arities may arise, but this is the only one we shall discuss here.) If three patches meet along edges meeting at a common vertex, we clearly can not use the same pair of parameters on all of them and expect the boundary parameters to correspond, so let the patches be $\underline{r} = \underline{f}(v,w)$, $\underline{r} = \underline{g}(w,u)$ and $\underline{r} = \underline{h}(u,v)$ respectively, with u, v and w each running from 0 to 1, and let the common boundaries be $v = u = 0$ between $\underline{f}$ and $\underline{g}$, and $v = w = 0$, $u = w = 0$ between the other two pairs of patches.

As we pass over the boundary from $\underline{f}$ to $\underline{g}$, it is natural to identify $\underline{f}(v,w)$ with $\underline{g}(w,-v)$, giving as conditions for C^0, C^1 and C^2 continuity:

$$\underline{f}(0,w) = \underline{g}(w,0), \tag{1}$$

$$(1) \text{ and } \underline{f}_v(0,w) + \underline{g}_u(w,0) = 0, \tag{2}$$

$$(1), (2) \text{ and } \underline{f}_{vv}(0,w) = \underline{g}_{uu}(w,0), \tag{3}$$

respectively. If condition (2) or (3) holds <u>mutatis</u> <u>mutandis</u> on all three common boundaries, however, we can deduce that

$$\underline{f}_v(0,0) = \underline{f}_w(0,0) = \ldots = \underline{h}_v(0,0) = 0, \tag{4}$$

so that the common vertex is a singular point of the surface. These conditions are therefore incapable of creating a smoothly-rounded corner.

Conditions (1), (2) and (3) assume, however, that lines of constant parameter must necessarily pass smoothly over patch boundaries, or that $\underline{r}$ is C^2 continuous in the parameters. This is an excessively strong condition merely to ensure that the surface is C^2. Suppose that in passing from $\underline{f}$ to $\underline{g}$ we instead identify $\underline{f}(v,w)$ with

$$\underline{g}(p(v,w), -q(v,w)),$$

where p and q are continuously differentiable scalar functions with

$$p(0,w) = w, \quad q(0,w) = 0, \quad q_v(0,w) = 1.$$

Then conditions (1), (2) and (3) are replaced by

$$\underline{f}(0,w) = \underline{g}(w,0), \tag{5}$$

$$(5) \text{ and } \underline{f}_v(0,w) = p_v(0,w)\, \underline{g}_w(w,0) - \underline{g}_u(w,0), \tag{6}$$

$$\begin{aligned}(5), (6) \text{ and } \underline{f}_{vv}(0,w) &= \underline{g}_{uu}(w,0) - 2p_v(0,w)\, \underline{g}_{wu}(w,0) \\ &+ p_v(0,w)^2\, \underline{g}_{ww}(w,0) + p_{vv}(0,w)\, \underline{g}_w(w,0) \\ &- q_{vv}(0,w)\, \underline{g}_u(w,0),\end{aligned} \tag{7}$$

respectively. Write $A(w) = -p_v(0,w)$ and $B(w) = -\frac{1}{2}q_{vv}(0,w)$, and choose p so that $p_{vv}(0,w) = A(w)\{A'(w) + B(w)\}$; then the conditions for C^0, C^1 and C^2 continuity can be expressed by saying that there exist A(w) and B(w) such that:

$$\underline{f}(0,w) = \underline{g}(w,0), \tag{8}$$

$$(8) \text{ and } \underline{f}_v(0,w) + \underline{g}_u(w,0) + A(w)\, \underline{g}_w(w,0) = 0, \tag{9}$$

$$\begin{aligned}(8), (9) \text{ and } &\underline{f}_{vv}(0,w) + A(w)\, \underline{f}_{vw}(0,w) + B(w)\, \underline{f}_v(0,w) \\ &= \underline{g}_{uu}(w,0) + A(w)\, \underline{g}_{wu}(w,0) + B(w)\, \underline{g}_u(w,0),\end{aligned} \tag{10}$$

respectively.

Let conditions similar to (8), (9) or (10), with the same functions A and B, hold between $\underline{g}$ and $\underline{h}$, and between $\underline{h}$ and $\underline{f}$. If the conditions are not to imply (4), we must in the C^1 case have

$$A(0) = 1, \tag{11}$$

and in the C^2 case

$$A(0) = 1, \quad B(0) = A'(0). \tag{12}$$

There are further conditions to be satisfied. In the situation illustrated in Figure 1, it is convenient to make $\underline{r}$ to be C^0, C^1 or C^2 -continuous in the parameters over most of the patch boundaries, or to satisfy the simpler conditions (1), (2) or (3) where the parameters change - reserving the weaker conditions (8), (9) or (10) for the three boundaries that meet at the exceptional corner vertex. In order that the other ends of these three boundaries should not be singular points, we find that we need the additional conditions

$$(C^1) \quad A(1) = A'(1) = 0, \tag{13}$$

or

$$(C^2) \quad A(1) = A'(1) = A''(1) = B(1) = B'(1) = B''(1) = 0. \tag{14}$$

If $\underline{f}$, $\underline{g}$ and $\underline{h}$ are to be polynomials, we must clearly take A and B to be at worst rational functions; for simplicity let them be polynomials of the smallest degree possible. Then for C^1 continuity we may take $A(w) = (1-w)^2$, and for C^2 continuity $A(w) = (1-w)^3$ and $B(w) = -3\,(1-w)^3$. Thus we are led to the conditions that each component of $\underline{f}$, $\underline{g}$ and $\underline{h}$ on the three adjoining patches is a linear combination of triads of polynomials $\{f(v,w),\ g(w,u),\ h(u,v)\}$ satisfying the continuity conditions:

$$(C^0) \quad f(0,w) = g(w,0),\ g(0,u) = h(u,0),\ h(0,v) = f(v,0); \tag{15}$$

$$(C^1) \quad f(0,w) = g(w,0),\ \text{etc.},$$
$$f_v(0,w) + g_u(w,0) + (1-w)^2\ g_w(w,0) = 0,\ \text{etc.}; \tag{16}$$

$$(C^2) \quad f(0,w) = g(w,0),\ \text{etc.},$$
$$f_v(0,w) + g_u(w,0) + (1-w)^3\ g_w(w,0) = 0,\ \text{etc.},$$
$$f_{vv}(0,w) + (1-w)^3\ \{f_{vw}(0,w) - 3\ f_v(0,w)\}$$
$$= g_{uu}(w,0) + (1-w)^3\ \{g_{wu}(w,0) - 3\ g_u(w,0)\},\ \text{etc..} \tag{17}$$

From this point onwards we shall restrict our attention to C^2 continuity, and shall suppose that in every patch other than the three with a common vertex the surface is represented by a bicubic polynomial (so that away from the corner we have a conventional bicubic spline). Then, to permit C^2 continuity along the boundaries $v = 1$, etc., we can restrict ourselves to solutions of (17) with the property that $f(1,w)$, $f_v(1,w)$, $f_{vv}(1,w)$, $f(v,1)$, $f_w(v,1)$, $f_{ww}(v,1)$, $g(1,u)$, $g_w(1,u)$, ..., $h_{vv}(u,1)$ are all polynomials of degree at most 3.

If f, g and h are allowed to be of degree at most d in each parameter, they will have this property if and only if they can be written in the form

$$f(v,w) = p_{3,3}(v,w) + (1-v)^3(1-w)^3\ p_{d-3,d-3}(v,w), \text{ etc.,} \qquad (18)$$

where $p_{r,r}$ denotes some polynomial of degree at most r in each of two variables.

Suppose that we take d = 4. Then there are precisely nine linearly independent triads of polynomials of the form (18) which satisfy conditions (17), namely:

f	g	h
1	1	1
$v^2\ w^2$	$w^2\ u^2$	$u^2\ v^2$
0	$w^2\ u^3$	$u^3\ v^2$
$v^3\ w^2$	0	$u^2\ v^3$
$v^2\ w^3$	$w^3\ u^2$	0
$v^3\ w^3$	0	0
0	$w^3\ u^3$	0
0	0	$u^3\ v^3$
$v^2+w^2-vw(1-v)^3(1-w)^3$	$w^2+u^2-wu(1-w)^3(1-u)^3$	$u^2+v^2-uv(1-u)^3(1-v)^3$

Any vector linear combination of these, unfortunately, we find must satisfy (4), so that the common vertex is still a singular point of the surface. [Worse, indeed, each of the three common boundaries degenerates to this point.] We must therefore go to a higher degree.

The process of finding linearly independent triads that satisfy (17) is in general made easier if we classify the triads according to their symmetry as follows:

(a) $f(w,v) = f(v,w)$, $g(w,u) = f(w,u)$, $h(u,v) = f(u,v)$;

(b) $f(w,v) = -f(v,w)$, $g(w,u) = f(w,u)$, $h(u,v) = f(u,v)$;

(c) $f(w,v) = \overline{f(v,w)}$, $g(w,u) = \omega f(w,u)$, $h(u,v) = \bar{\omega} f(u,v)$;

(d) $f(w,v) = -\overline{f(v,w)}$, $g(w,u) = \omega f(w,u)$, $h(u,v) = \bar{\omega} f(u,v)$.

Here $\omega = \exp(\pm 2i\pi/3)$, and $\bar{\ }$ denotes the complex conjugate.

Conditions (17) then reduce to conditions on f alone as follows:

(a) $2\ f_v(0,w) + (1-w)^3\ f_w(0,w) = 0;$ (19a)

(b) $f(0,w) = 0$, and

$$f_{vv}(0,w) + (1-w)^3\{f_{vw}(0,w) - 3\ f_v(0,w)\} = 0; \qquad (19b)$$

(c) $\omega\,[2\ f_v(0,w) + (1-w)^3\ f_w(0,w)]$ is pure imaginary, and $\omega f(0,w)$ and $\omega\,[f_{vv}(0,w) + (1-w)^3\{f_{vw}(0,w) - 3\ f_v(0,w)\}]$ are real; (19c)

(d) $\omega\,[2\ f_v(0,w) + (1-w)^3\ f_w(0,w)]$ is real, and $\omega\ f(0,v)$ and $\omega[f_{vv}(0,w) + (1-w)^3\{f\ \ (0,w) - 3\ f_v(0,w)\}]$ are pure imaginary. (19d)

With the help of this device, it can be found that there are thirteen linearly independent triads for $d = 5$, and twenty-four for $d = 6$. Although (4) no longer necessarily holds, it will be found that when $d = 5$ it is always the case that $f(1,0) = f(0,1)$, so that the surface will have a triple point. We must therefore go up to polynomials of degree 6 before our basis contains enough triads of functions to represent a simple rounded contour.

To show that $d = 6$ is sufficient, we have computed an approximation to the octant of a sphere by three such quasi-rectangular patches, in each of which $\underline{r}$ is a polynomial of degree 6 in two parameters, made up out of the 24 basis polynomials. Figure 2 shows a perspective view of the intersections of the patches with a set of parallel planes, and Figure 3 shows the lines of constant parameter on the three patches. In the neighbourhood of the common vertex it will be seen that the parameters tend to change rapidly, whereas the surface (which has actually been forced to have the same symmetries as the octant itself) is apparently uniformly smooth.

This work owes much to the support of the Cowley Body Plant of B.L. Cars Ltd (Oxford), and to discussions with Mr T. Stacey. A brief account appeared in [1].

[1] Handscomb, D.C.: Turning corners with four-sided polynomial patches. In Cheney, E.W.(ed.): Approximation Theory III, New York, Academic Press, 1980, 481-484.

Oxford University Computing Laboratory
19 Parks Road
Oxford OX1 3PL
England

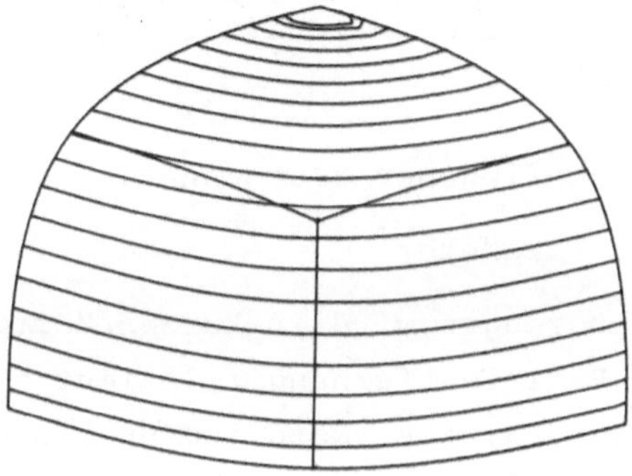

Figure 2. Approximation to octant of sphere, plane sections.

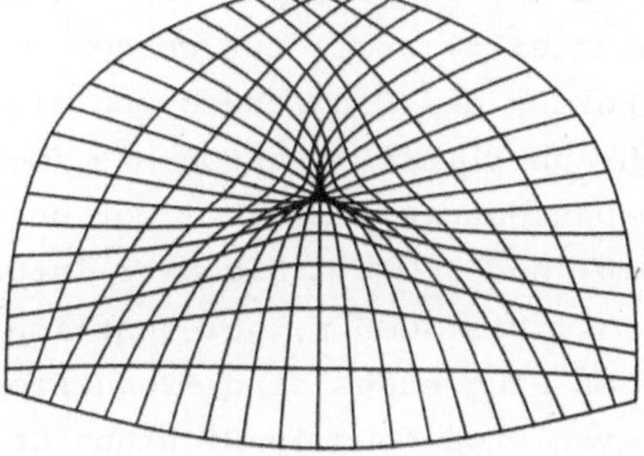

Figure 3. Approximation to octant of a sphere, parameter lines.

EINIGE APPROXIMATIONEN VON FLUGBAHNEN

K. Nixdorff

This paper deals with the approximation of courses flown by aircraft, with the extrapolation of the following courses of aircraft and with the approximation of tables which are used in connection with the courses of aircraft for surface-to-air and air-to-surface problems.

1. Approximation bereits geflogener Flugbahnen.

1.1 Aufgabenstellung und Parameter.

Der Standort eines Flugzeuges wird in kurzen Zeitabständen Δt mit einem Radar nach Radialentfernung r, Azimut α (bezüglich einer gewählten Grundrichtung, z.B. Nordrichtung) und Elevation ε vermessen. Mit diesen fehlerbehafteten Werten ist die Flugbahn des Flugzeuges zu ermitteln.

Die Dauer der Meßintervalle Δt reicht meist von 10 ms bis 50 ms. Der Fehler der Zeitmessung wird vernachlässigt. Die hier interessierende Radialentfernung r reicht bis 4000 m, gelegentlich weiter. Als mittlerer Fehler der Radialentfernung kann für alle Entfernungen 1 m angenommen werden. Das Azimut kann jeden Wert zwischen 0° und 360°, die Elevation kann jeden Wert zwischen etwa 0° und 90° annehmen. Die mittleren Fehler der beiden Winkelmessungen dürfen zu 1^- (gesprochen: ein Strich: der Vollkreis wird in 6400^- unterteilt) angesetzt werden.

Als Bahngeschwindigkeit v des Flugzeuges muß ein Bereich von 0 m/s (hovernder Hubschrauber) bis etwa 600 m/s (moderner Jet) berücksichtigt werden. Das Hauptaugenmerk liegt auf Bahngeschwindigkeiten zwischen 200 m/s und 400 m/s. Die hier interessierende Flughöhe h reicht von 15 m bis 4000 m, gelegentlich höher.

Für das Schlingern des Flugzeuges sind Frequenzen f um etwa 0,1 Hz und Amplituden a von etwa 15 m in Richtung der Flugbahnnormalen und -binormalen typisch.

Die oben angegebenen Parameter unterliegen verschiedenen Beschränkungen.

Mit der Flughöhe nimmt auch die maximale Bahngeschwindigkeit ab. Für die maximale Beschleunigung und für die maximale Verzögerung bestehen flugzeugtypabhängige Schranken. Überdies wird der Pilot zu hohe Radialbeschleunigungen b_r, etwa oberhalb 5 g (g bedeutet den üblichen Wert der Fallbeschleunigung) vermeiden.

Im allgemeinen will der Pilot so fliegen, daß sein Treibstoffverbrauch minimal wird. Dies entspricht etwa der Forderung

$$\int_s b_t \, ds = \text{Min},$$

wobei b_t die Tangentialbeschleunigung und ds das Wegelement der Flugbahn s bedeuten.

Manche Flugmanöver sollen ohne Rücksicht auf Treibstoffverbrauch in möglichst kurzer Zeit ausgeführt werden.

1.2 Numerisches Vorgehen

Ursprünglich wurde durch die vermessenen Flugzeugorte eine Interpolations- oder eine Ausgleichsgerade gelegt. Versuche, die Genauigkeiten zu steigern, führten zur Verwendung von Polynomen mit der Flugzeit t als unabhängige Veränderliche für dann meist rechtwinklig-cartesisch gewählte Koordinaten des Ortsvektors $\vec{r}$ der Flugzeugbahn, z.B. bei U. Simon [1] , oder zu anderen nichtlinearen Funktionen, z.B. bei K. Nixdorff und E. Schmitt [2]. Es zeigt sich, daß so trotz erheblichem Aufwand die Genauigkeit nicht genügend erhöht werden kann, und zwar gleichgültig, ob das angesetzte Polynom als Taylorentwicklung, als Interpolationspolynom oder als Ausgleichspolynom aufgefaßt wurde. Dies liegt vermutlich daran, daß sich bei diesem Vorgehen das Schlingern wie eine (erhebliche) Vergrößerung der Meßfehler auswirkt.

Deshalb wird versucht, durch den Ansatz

$$r(t) = r_1(t) + r_2(t)$$

mit

$$r_2(t+f^{-1}) = r_2(t)$$

den Einfluß des Schlingerns abzuspalten.

Da die Flugbahn r(t) bereits nach einer Beobachtungszeit approximiert werden soll, die kürzer ist als die Schwingungsdauer

f^{-1} des Schlingerns $r_2(t)$, können die Parameter des Schlingerns so nicht genügend genau aus den vermessenen Flugzeugorten gewonnen werden.

Deshalb wird versucht, die vom Piloten angestrebte Flugbahn $r_1(t)$ zum Ausgleichen der vermessenen Flugzeugorte vorzugeben. Ist diese Flugbahn nicht bekannt, so sind mehrere, im Rechner gespeicherte Flugbahnen (genauer: mit noch zu bestimmenden Parametern versehene Flugbahnprofile) vorzugeben und diejenige mit dem kleinsten Fehler ist weiterzuverwenden. "Kleinster Fehler" ist dabei zu präzisieren, insbesondere ist zu untersuchen, ob und wie stark Meßwerte nach ihrem Alter zu wichten sind. Ist die geflogene Flugbahn nicht mit einem der gespeicherten Flugbahnprofile genügend genau erfaßbar, so sollte diese Flugbahn automatisch zur Speicherung eines neuen entsprechenden Flugbahnprofiles führen.

Das systematische Erraten der angestrebten Flugbahn $r_1(t)$ wird dadurch erleichtert, daß im allgemeinen bei Berücksichtigung der mutmaßlichen Absichten des Piloten nur wenige typische Flugbahnprofile (Standardflugbahnprofile) in Frage kommen. Solche Flugbahnprofile sind z.B. gerader Vorbeiflug, gerader Überflug, Umfliegen eines Geländestückes, Umkreisen eines Geländestückes, der sogenannte Sprung vorwärts, der sogenannte Sprung rückwärts.

Außer dem letzten Beispiel sind die aufgeführten Flugbahnprofile ebene Kurven. Dies kann als typisch angesehen werden. Der Sprung seitwärts verläuft, wie viele zunächst nicht ebenen Flugbahnprofile, genügend genau in der Mantelfläche eines gedachten, lotrecht stehenden Zylinders und ist durch Abwickeln dieses Zylinders in eine ebene Kurve überführbar.

Fast alle Flugbahnprofile sind nicht durch eine (vektorwertige) Funktion beschreibbar. Versuche, Polynomsplines zu verwenden, führten zu unsinnigem Verlauf der Radialbeschleunigung. Wie bei H.B. Fallmeier [3] gezeigt wurde, wird der Verlauf der Flugbahn besser beschrieben, wenn Stücke von Geraden, Kreisbögen und Klothoiden zum Splinen verwendet werden.

Als Beispiel seien von H.B. Fallmeier [3, Abb. 61-67] erhaltene Kurven für den Sprung seitwärts wiedergegeben.

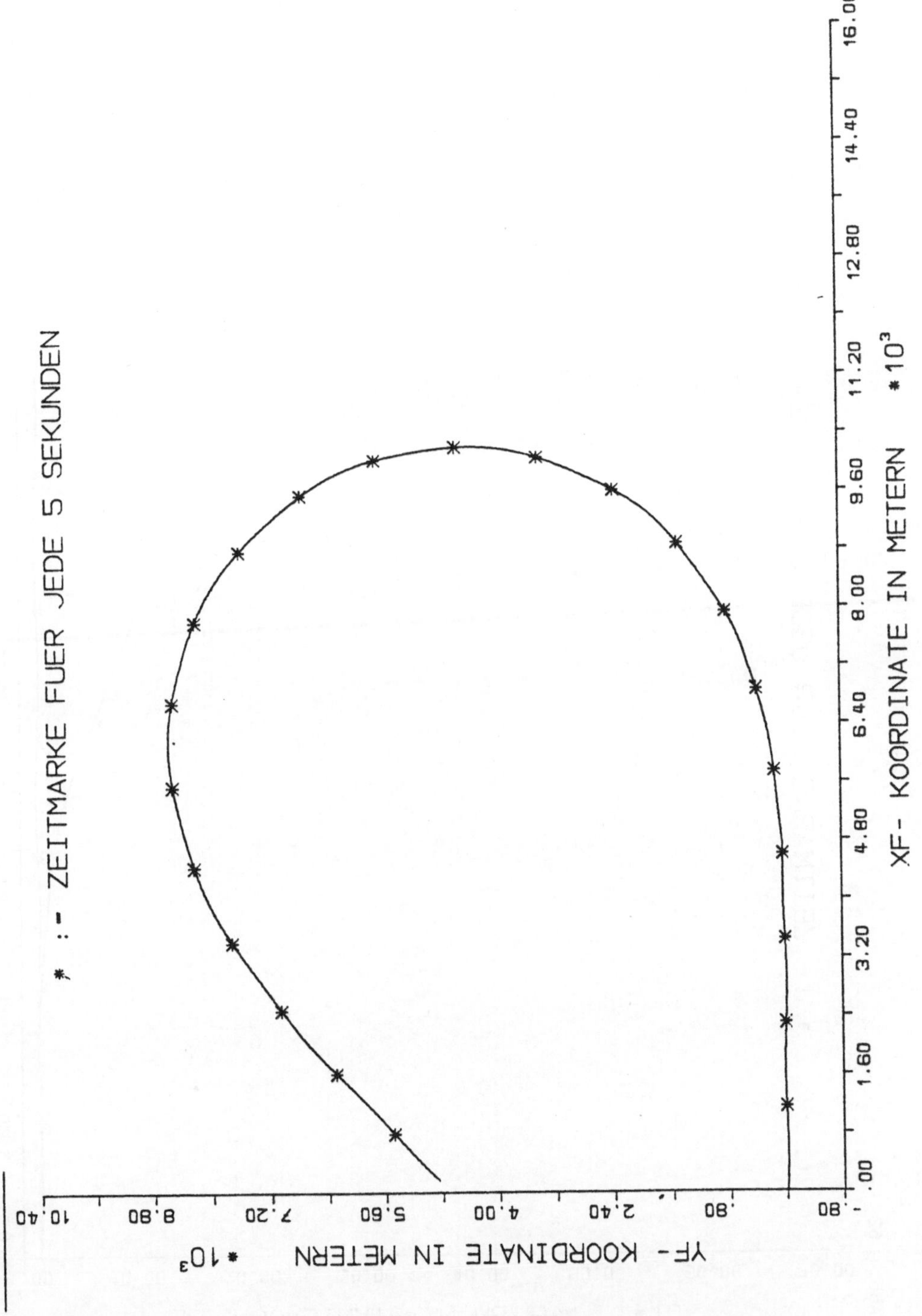
* :- ZEITMARKE FUER JEDE 5 SEKUNDEN
XF- KOORDINATE IN METERN *10³
YF- KOORDINATE IN METERN *10³
.00
1.60
3.20
4.80
6.40
8.00
9.60
11.20
12.80
14.40
16.00
-.80
.80
2.40
4.00
5.60
7.20
8.80
10.40

Abb.: 61

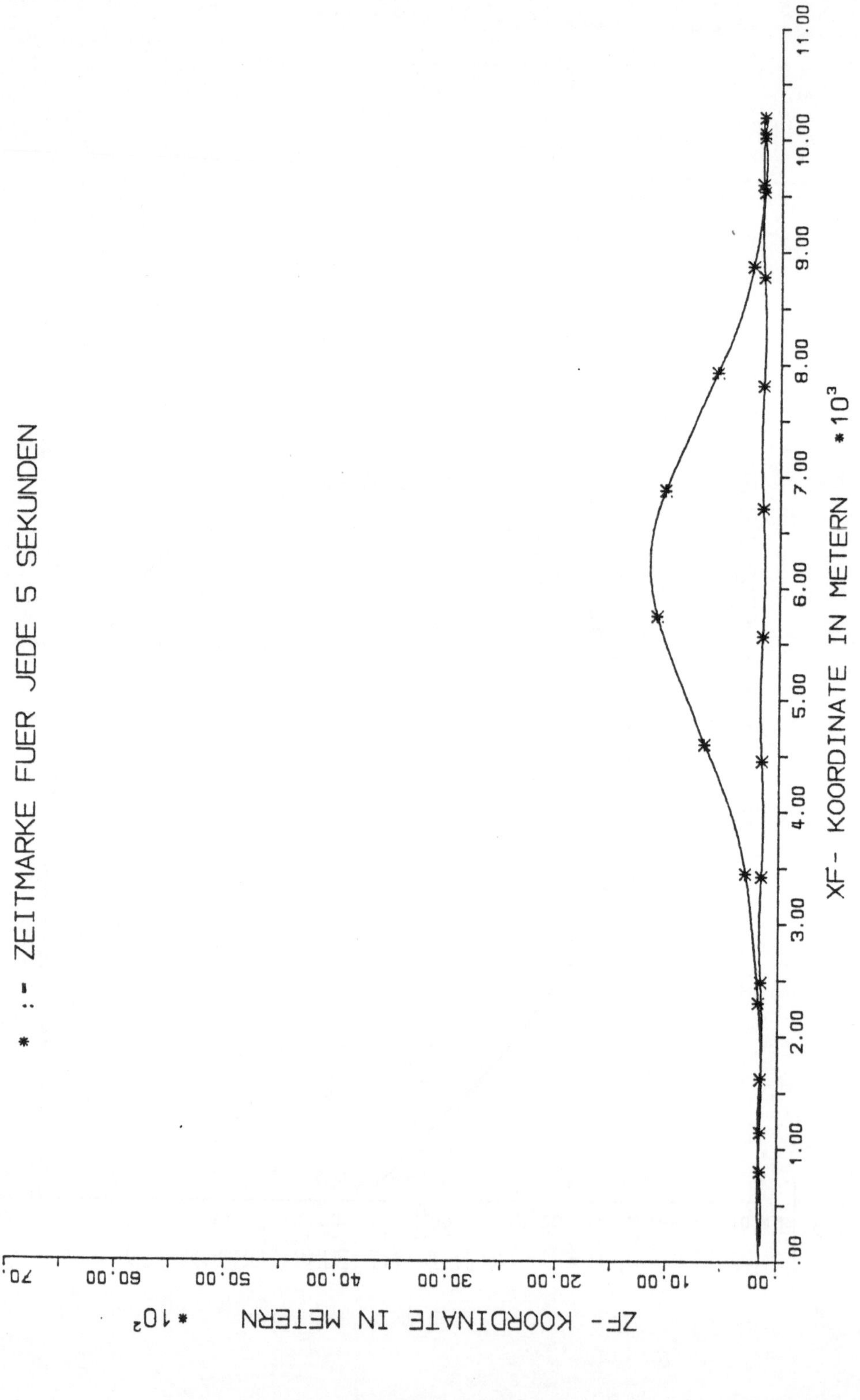

* :- ZEITMARKE FUER JEDE 5 SEKUNDEN
XF- KOORDINATE IN METERN *10³
ZF- KOORDINATE IN METERN *10²
.00
1.00
2.00
3.00
4.00
5.00
6.00
7.00
8.00
9.00
10.00
11.00
10.00
20.00
30.00
40.00
50.00
60.00
70.00

Abb.: 62

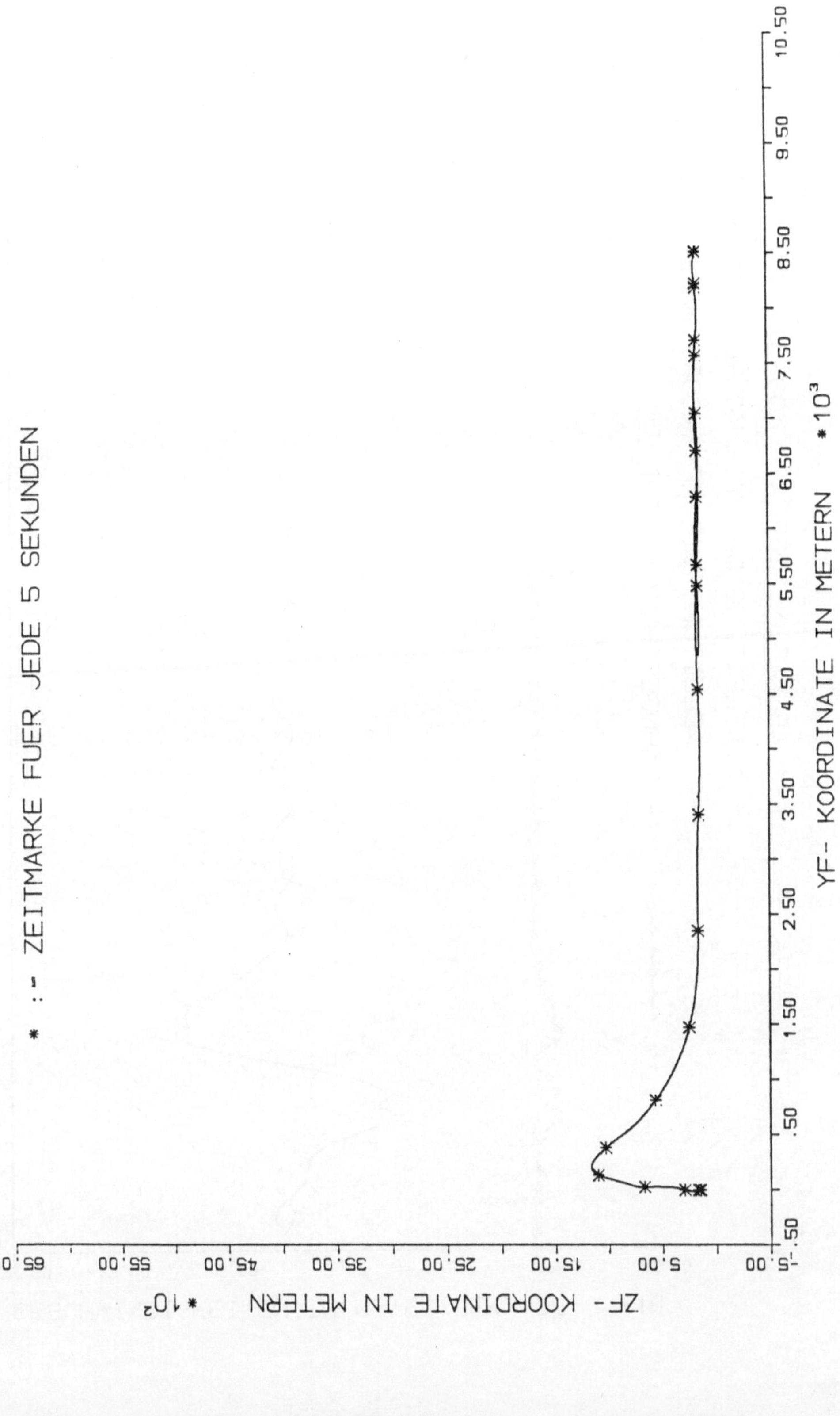
* := ZEITMARKE FUER JEDE 5 SEKUNDEN
YF- KOORDINATE IN METERN *10^3
ZF- KOORDINATE IN METERN *10^2
-.50
.50
1.50
2.50
3.50
4.50
5.50
6.50
7.50
8.50
9.50
10.50
-5.00
5.00
15.00
25.00
35.00
45.00
55.00
65.00

Abb.: 63

Abb.: 64

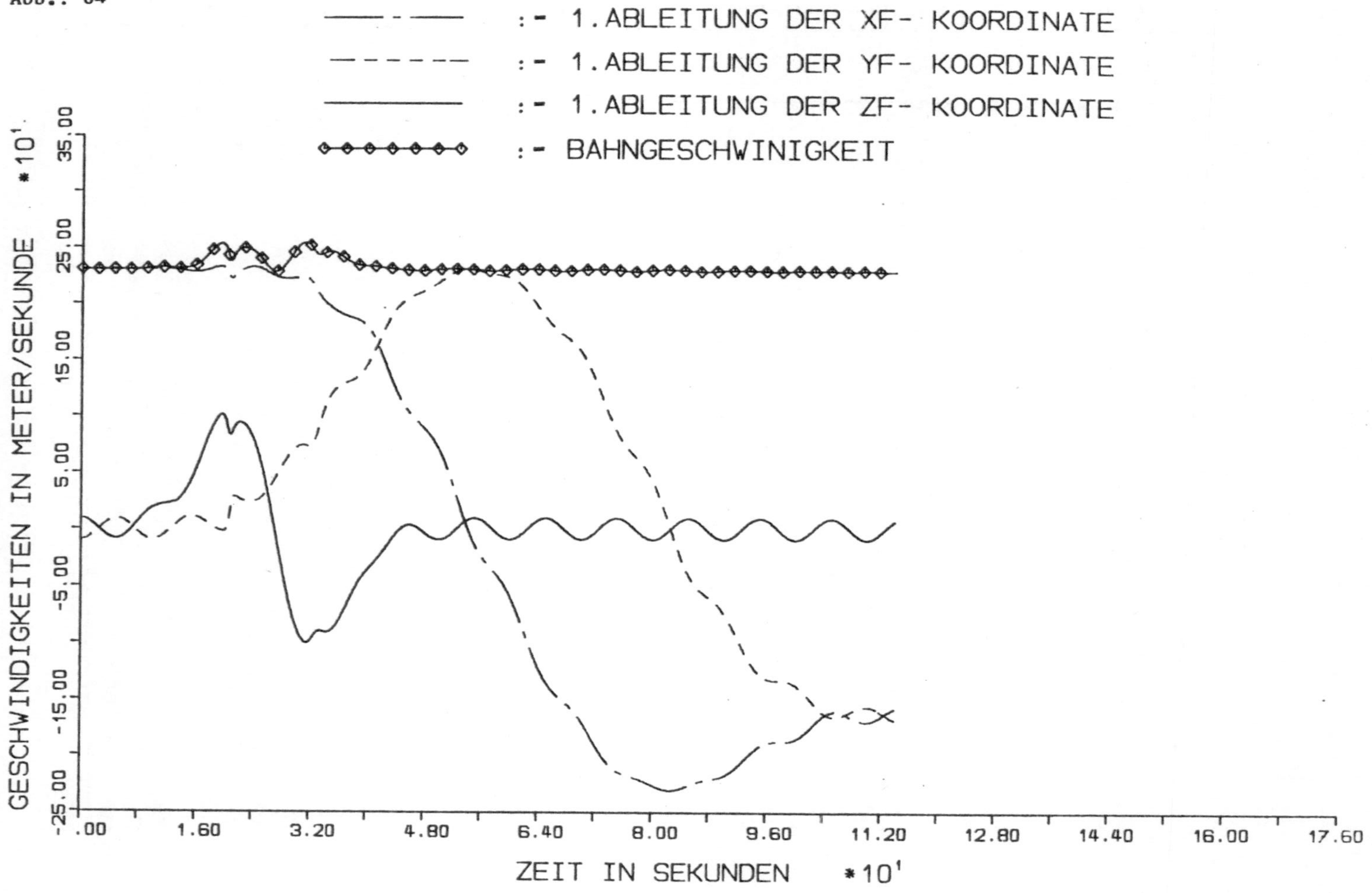
:- 1.ABLEITUNG DER XF- KOORDINATE
:- 1.ABLEITUNG DER YF- KOORDINATE
:- 1.ABLEITUNG DER ZF- KOORDINATE
:- BAHNGESCHWINIGKEIT
GESCHWINDIGKEITEN IN METER/SEKUNDE *10¹
35.00
25.00
15.00
5.00
-5.00
-15.00
-25.00
.00
1.60
3.20
4.80
6.40
8.00
9.60
11.20
12.80
14.40
16.00
17.60
ZEIT IN SEKUNDEN *10¹

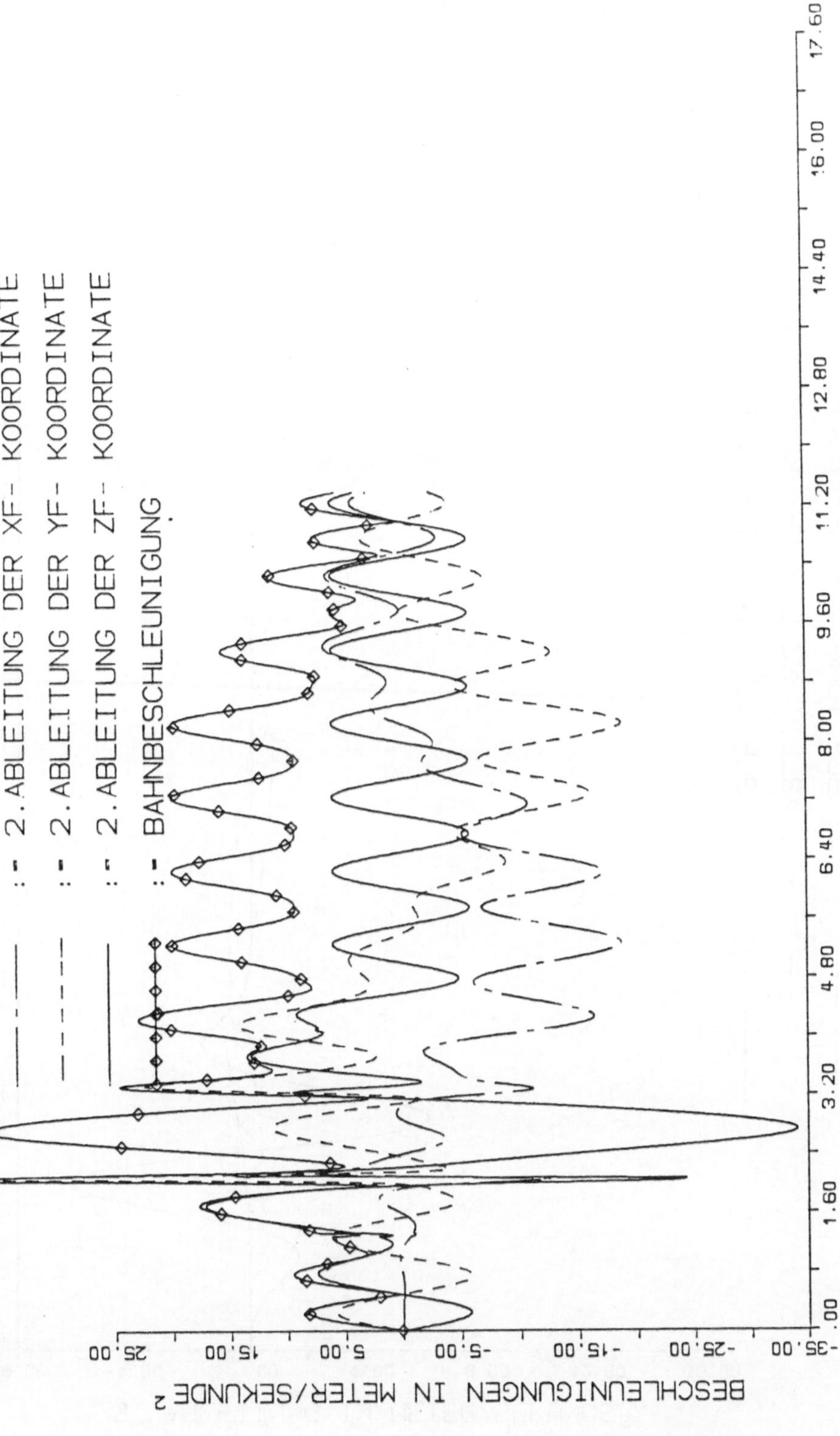
: 2. ABLEITUNG DER XF- KOORDINATE
: 2. ABLEITUNG DER YF- KOORDINATE
: 2. ABLEITUNG DER ZF- KOORDINATE
: BAHNBESCHLEUNIGUNG
ZEIT IN SEKUNDEN *10[1]
.00
1.60
3.20
4.80
6.40
8.00
9.60
11.20
12.80
14.40
16.00
17.60
BESCHLEUNIGUNGEN IN METER/SEKUNDE 2
25.00
15.00
5.00
-5.00
-15.00
-25.00
-35.00

Abb.: 65

Abb.: 66

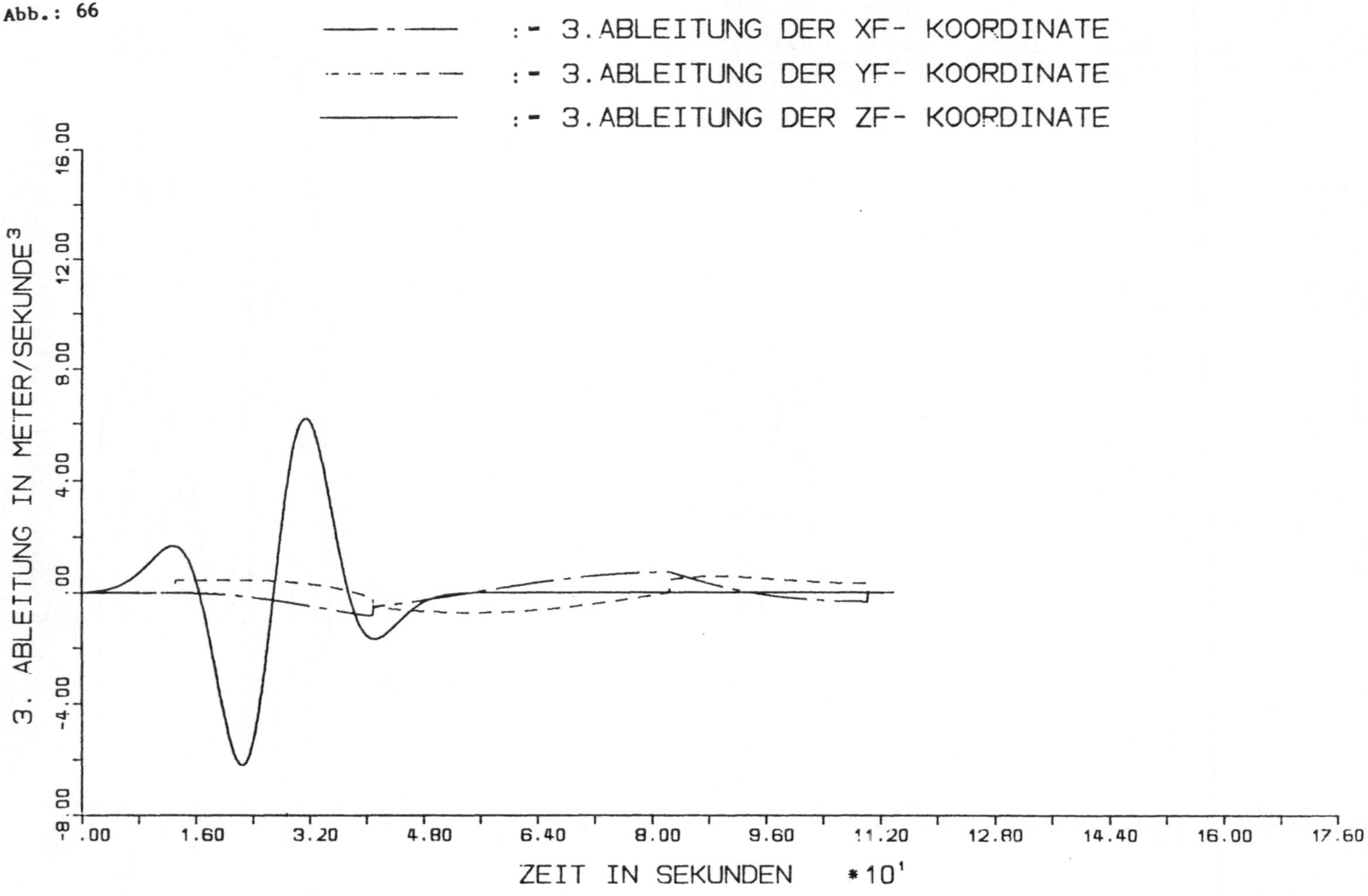
:- 3.ABLEITUNG DER XF- KOORDINATE
:- 3.ABLEITUNG DER YF- KOORDINATE
:- 3.ABLEITUNG DER ZF- KOORDINATE
3. ABLEITUNG IN METER/SEKUNDE^3
16.00
12.00
8.00
4.00
.00
-4.00
-8.00
.00
1.60
3.20
4.80
6.40
8.00
9.60
11.20
12.80
14.40
16.00
17.60
ZEIT IN SEKUNDEN *10^1

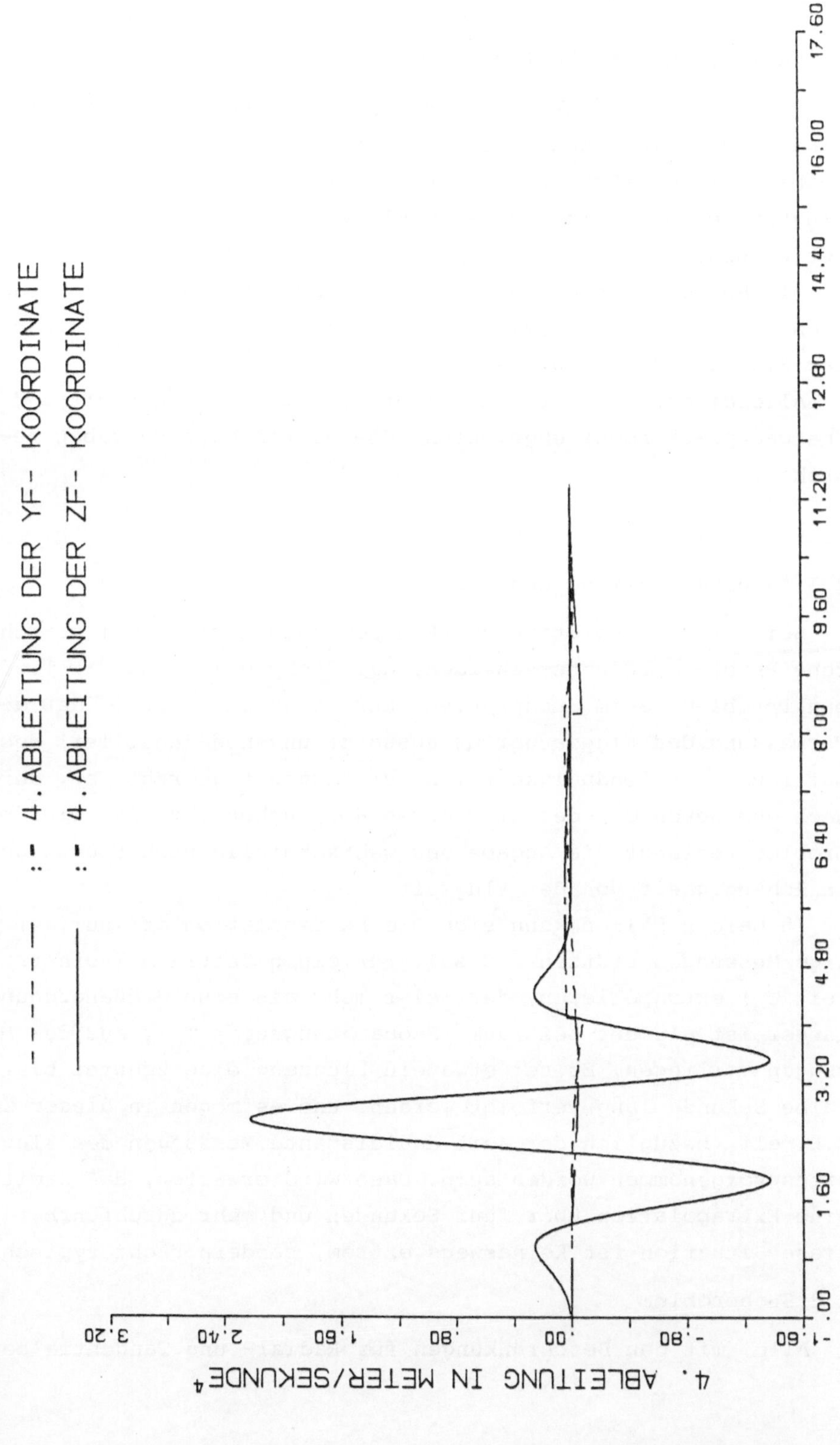
4. ABLEITUNG DER XF- KOORDINATE
4. ABLEITUNG DER YF- KOORDINATE
4. ABLEITUNG DER ZF- KOORDINATE
4. ABLEITUNG IN METER/SEKUNDE⁴
ZEIT IN SEKUNDEN *10¹
-1.60
-.80
.00
.80
1.60
2.40
3.20
.00
1.60
3.20
4.80
6.40
8.00
9.60
11.20
12.80
14.40
16.00
17.60

Abb.: 67

1.3 Weitere Möglichkeiten.

Radialgeschwindigkeit $\dot{r}_1$ und Tangentialgeschwindigkeit $\dot{r}_t$ sind unabhängig voneinander und von den Ortskoordinaten r,α,ε meßbar. Diese Meßdaten könnten zur Erhöhung der Genauigkeit der Flugbahnapproximation, etwa im Sinne der Kalmanfilter, verwendet werden.

Die Beschränkungen für Bahn- und Radialbeschleunigung lassen sich statt zur nachträglichen Kontrolle sofort in das Ermitteln der Flugbahn einbauen.

Allerdings ist stets zu beachten, daß für Rechenzeit und Speicherplatz recht enge, nicht überschreitbare Vorgaben bestehen.

2. Extrapolation von Flugbahnen.

2.1 Aufgabenstellung und Parameter.

Bei der Extrapolation von Flugzeugbahnen sind zwei verschiedene Probleme zu unterscheiden, das Suchproblem und das Treffpunktproblem. Beim Suchproblem handelt es sich darum, daß die Vermessung des Flugzeugortes aussetzt und nun jener Teil des Luftraumes in Abhängigkeit von der Flugzeit gefragt ist, der nach dem neuen Ort des Flugzeuges abzusuchen ist. Das Treffpunktproblem verlangt die Angabe des wahrscheinlichsten Flugzeugortes in Abhängigkeit von der Flugzeit.

In beiden Fällen kann sich die Extrapolation oft nur auf wenige Messungen stützen und soll für einen Zeitraum (Vorhersagezeit T_V) extrapolieren, der meist mehr als eine Größenordnung länger ist als der Zeitraum (Beobachtungszeit T_B), aus dem Messungen vorliegen. Es sei etwa ein Flugzeug eine zehntel bis eine halbe Sekunde lang verfolgt worden, und es mögen in dieser Zeit z.B. elf bezüglich der Zeit äquidistante Messungen des Flugzeugortes vorgenommen worden sein. Dann wird erwartet, daß damit eine Extrapolation über fünf Sekunden und mehr durchführbar ist. Diese Situation ist keineswegs extrem, sondern recht typisch.

2.2 Suchproblem

Alle, mit den Beschränkungen für Radial- und Tangentialbeschleunigungen verträglichen, am letzten vermessenen Flugzeugort anschließenden möglichen Flugbahnen erfüllen im Raum einen sich stark öffnenden Trichter (sicherer Trichter), der noch im Verlauf der üblicherweise gewünschten Vorhersagezeiten zu groß wird.

Allerdings enthält dieser Trichter sicher das Flugzeug (sofern es noch fliegt). Werden nicht alle physikalisch möglichen Flugbahnen betrachtet, sondern nur jene, die mit dem bisher geflogenen Flugbahnprofil verträglich sind, so wird innerhalb des sicheren Trichters ein kleinerer Trichter erhalten, der wegen der angenommenen Konstanz des Flugbahnprofils, also der (vermutlichen) Absicht des Piloten, nicht mehr sicher das Flugzeug enthält (unsicherer Trichter).

Sicherer bzw. unsicherer Trichter können durch weitere Informationen verkleinert werden, etwa durch Kenntnisse bzw. Mutmaßungen über die Absichten des Piloten.

2.3 Treffpunktproblem.

Die Vorhersage beim Treffpunktproblem erfolgt im wesentlichen dadurch, daß die Flugbahnapproximation aus der Beobachtungszeit in die Vorhersagezeit verlängert wird. Können nach Abgabe einer Vorhersage Flugzeugorte vermessen werden, was in der Praxis beim Treffpunktproblem die Regel ist, so kann aus einem Vergleich des Vorhergesagten mit dem vermessenen Flugzeugort eine Verbesserung der nächsten Vorhersage versucht werden.

Die Optimierung der Flugbahnapproximation während der Beobachtungszeit im Hinblick auf die Genauigkeit der Vorhersage wurde z.B. von K. Nixdorff und E. Schmitt [2] und U. Simon [1] untersucht. Es zeigt sich, daß der Polynomansatz (mit für die ganze Beobachtungszeit konstanten Koeffizienten $\vec{r}_o$ bis $\vec{c}_n$)

$$\vec{r}(t) = \vec{r}_o + \vec{v}_o\, t + \frac{1}{2}\vec{b}_o\, t^2 + \ldots + \vec{c}_n\, t^n$$

für die Flugbahn vor allem infolge des "Weglaufens" der Polynome einen optimalen Grad n hat, der je nach Größe der Meßfehler und des Schlingerns und je nach Flugbahn und bei der Rechnung verwendeter Ziffernlänge gelegentlich bereits bei eins, meist bei zwei oder drei liegt.

Diese Erscheinung tritt auf, wenn die der Messung zugrunde liegenden Kugelkoordinaten r, α, ε in der Rechnung entsprechend dem obigen Ansatz im Ortsvektor polynomial oder für jede Koordinate einzeln polynomial approximiert weiterverwendet werden und auch, wenn zur Approximation die gemessenen Kugelkoordinaten sofort in rechtwinklig-cartesische Koordinaten transformiert werden und zwar sowohl bei Taylorentwicklung, als auch beim Interpolieren und beim Ausgleichen.

Die Verwendung der Kugelkoordinaten ist anscheinend vorteilhaft, weil die Rechenzeit für die Koordinatentransformation entfällt. Tatsächlich treten bei Beibehaltung des polynomialen Ansatzes für den Ortsvektor rechenzeitaufwendige Winkelfunktionen auf, so daß die Rechenzeit ansteigen kann. Auch können sich bei polynomialem Ansatz für jede einzelne Kugelkoordinate unbefriedigende Bahnverläufe ergeben.

Bei dieser Gelegenheit sei auf die Begriffe "lineare Extrapolation" und "nichtlineare Extrapolation" mit deren Sonderfall "quadratische Extrapolation" eingegangen. Diese Begriffe werden in drei verschiedenen Bedeutungen verwendet. Sie können sich beziehen auf das räumliche Aussehen der Flugbahn (wobei dann "quadratisch" soviel wie "Kegelschnitt" bedeutet) oder auf die zeitliche Abhängigkeit des Ortsvektors (und damit der rechtwinklig-cartesischen Koordinaten) oder auf die zeitliche Abhängigkeit der gerade verwendeten Ortskoordinaten (etwa der Kugelkoordinaten r, α, ε). Offensichtlich können hier erhebliche Mißverständnisse auftreten.

Der Vergleich bisher vorhergesagter und bereits vermessener Ortskoordinaten kann in sinngemäßer Anwendung der für Analogregler üblichen Regelvorschriften zur Verbesserung der nächsten Vorhersage benutzt werden. Bedeuten $\vec{r}_r^{\,(j)}(t)$ der zunächst im j-ten Zeitpunkt seit Beginn der Vorhersagen roh für den Zeitpunkt t vorhergesagte Ortsvektor, $\vec{r}_v^{\,(j)}(t)$ der entsprechende verbesserte Ortsvektor und $\vec{r}(t_k)$ der im Zeitpunkt t_k gemessene Ortsvektor, so gilt mit den Koeffizienten c_p, c_d und c_i und der (konstanten) Dauer Δt der Meßintervalle

$$\vec{r}_v^{\,(j)}(t) = \vec{r}_r^{\,(j)}(t) + c_p[\vec{r}(t_j)-\vec{r}_v^{\,(j-1)}(t_j)] + $$
$$+ \frac{c_d}{\Delta t}[\vec{r}(t_j)-\vec{r}_v^{\,(j-1)}(t_j)-\vec{r}(t_{j-1})+\vec{r}_v^{\,(j-2)}(t_{j-1})] + $$
$$+ c_i\ \Delta t \sum_{k=1}^{j} [\vec{r}(t_k)-\vec{r}_v^{\,(k-1)}(t_k)].$$

Diese Verbesserungsvorschrift wurde von K. Nixdorff und E. Schmitt [2], von D. Topler [4] und von H.B. Fallmeier [5] getestet. Es zeigte sich, daß der Einfluß des Koeffizienten c_d mit der Zeit zunehmend den Unterschied zwischen den Ortsvektoren $\vec{r}_v^{\,(j)}(t > t_j)$ und $\vec{r}(t > t_j)$ vergrößert und daß der Einfluß des Koeffizienten c_i entgegengesetzt wirkt. Die besten Ergebnisse wurden mit den Kombinationen $c_p \neq 0$, $c_d \neq 0$, $c_i \neq 0$ und $c_p \neq 0$, $c_d = 0$, $c_i \neq 0$ erzielt.

Dabei wurden die Koeffizienten c_p, c_d und c_i für die ganze Flugbahn konstant gehalten. Überlegungen von D. Topler [4] zur Wahl der Koeffizienten zeigen, daß diese Koeffizienten mit Veränderung der Flugbahn veränderlich sein sollten.

Grundsätzlich liefert die Verwendung von Standardflugbahnprofilen recht genaue Vorhersagen, sofern tatsächlich ein Standardflugbahnprofil geflogen und das Standardflugbahnprofil richtig erraten wurde. In der Praxis kann das Speichern genügend vieler Standardflugbahnprofile zu speicheraufwendig und die Auswahl des günstigsten Standardflugbahnprofiles zu rechenzeitaufwendig sein.

3. Approximation von Boden-Luft- und Luft-Boden-Tafeln.

3.1 Aufgabenstellung und Parameter.

Im Zusammenhang mit den Flugbahnen von Flugzeugen stehen die Boden-Luft- und die Luft-Boden-Tafeln. Die Boden-Luft-Tafeln haben zwei Eingangsparameter (Schrägentfernung r und Elevation ε des Flugzeuges von der Bodenstation aus gesehen); die Luft-Boden-Tafeln haben drei Eingangsparameter (Winkel φ zwischen Horizontalebene und Richtung der Fluggeschwindigkeit gegen umgebende Luft, Flughöhe über Grund (Erdboden), Betrag v_{TAS} der Fluggeschwindigkeit gegen umgebende Luft). Die Tafeln sind mit so kleinen Schrittweiten der Eingangsparameter aufgestellt worden, daß in den Tafeln stets linear interpoliert werden kann. Daraus folgt, daß an den Stützstellen selbst die Tafelwerte etwas genauer sind, als in der Praxis benötigt.

Verschiedene Prozeßrechner benötigen die in den Tafeln enthaltenen Informationen und dies mit Zugriffszeiten im Mikrosekundenbereich. Die Information muß daher im Kernspeicher des Rechners abrufbar sein. Die Tafeln selber sind aber zu umfangreich für den Kernspeicher und werden deshalb durch Polynome (wegen deren kurzer Rechenzeit) ersetzt. Die Koeffizienten der Polynome sind so zu wählen, daß bei ausreichender Genauigkeit und hinreichend kleinem Speicherbedarf die Rechenzeit möglichst klein wird.

3.2 Ausgleichspolynome mit Überlappung.

In solchen Fällen wurden die Tafeln in (möglichst) sich überlappende Bereiche unterteilt und in jedem Bereich durch ein Polynom mit konstanten Koeffizienten und allen Eingangsparametern als unabhängigen Veränderlichen ersetzt.

Die Überlappung sollte soweit gehen, daß bei den oft mehrere ineinandergeschachtelte Iterationen enthaltenden Rechnungen (siehe z.B. K. Nixdorff und E. Schmitt [6]) die Bereiche nicht gewechselt werden müssen. Ist so ein Wechsel doch erforderlich, so muß überprüft werden, ob nicht dadurch die Durchführung der Rechnungen zu sehr erschwert wird. In ungünstigen Fällen kommen Iterationen nicht mehr zum Stehen.

Die Koeffizienten werden durch Ausgleichen, meist nach Gauß, manchmal nach Tschebytscheff, bestimmt. Es wird bei jedem Polynom versucht, die Anzahl der nichtverschwindenden Koeffizienten möglichst klein zu halten. Je nach Verlauf der Tafelwerte kann es dabei sinnvoll sein, von vornherein Koeffizienten zu Null zu wählen.

Der notwendige Umfang der Überlappung ist oft nicht ausreichend sicher feststellbar. Auch wird durch die Überlappung die Anzahl der zu speichernden Koeffizienten vermehrt. Deshalb wird auf die Überlappung völlig verzichtet und auf Polynomspline übergegangen.

3.3 Ausgleichspolynomspline.

Wird für jede Tafel sofort ein Polynomspline in allen Eingangsparametern angesetzt, so ergeben sich häufig wegen der geforderten Genauigkeit zu viele Koeffizienten. S. Kempfle [7] schlug deshalb ein anderes Verfahren vor, das mit sehr viel weniger Koeffizienten auskommt, wie von P. Schackmar [8] an umfangreichen und typischen Boden-Luft-Tafeln gezeigt wurde.

Das Vorgehen nach S. Kempfle sei der Einfachheit halber an Hand einer Boden-Luft-Tafel erläutert.

Es werden sämtliche Zeilen oder, was zu erproben ist, Spalten durch Polynomspline gleichen Grades und gleicher Anschlußstellen ersetzt. Für die Koeffizienten dieser Polynomspline, die zu gleichen Potenzen gehören, werden wiederum Polynomspline angesetzt.

Zusammen mit einer für jede Tafel neu vorgenommenen Anpassung des Grades der Splinepolynome und der Wahl der von vornherein Null zu setzenden Koeffizienten konnten Speicherbedarf und Rechenzeit erheblich verkürzt werden.

3.4 Weitere Möglichkeiten.

An Stelle einer ursprünglichen Tafel kann auch eine durch Verwendung größerer Schrittweiten verkürzte Hilfstafel zusammen mit einer der Verkürzung entsprechenden aufwendigen Interpolationsvorschrift im Rechner gespeichert werden. Dieses Vorgehen scheint im allgemeinen umso konkurrenzfähiger zu sein, je mehr Eingangsparameter die Tafel hat.

Eine andere Möglichkeit wurde bei einem bereits eingeführten Rechner, freilich für Boden-Boden-Probleme, verwirklicht. Hier werden die gewünschten Tafelinformationen bei jeder Rechnung durch numerische Integration von gewöhnlichen Differentialgleichungen angenähert erhalten, und diese Näherung wird anschliessend durch einfache Korrekturformeln zur gewünschten Genauigkeit verbessert.

Schließlich sei noch daran erinnert, daß früher in der Praxis für Aufgaben dieser Art keine Digital-, sondern Analogrechner verwendet wurden. Möglicherweise wäre ein auf die jeweilige Aufgabe hin entwickelter Hybridrechner eine befriedigende Lösung.

LITERATURVERZEICHNIS:

[1] U. Simon: Approximation von Zielflugbahnen, Studienarbeit an der Hochschule der Bundeswehr Hamburg, 1979.

[2] K. Nixdorff und E. Schmitt: Extrapolation fehlerbehafteter Messungen, Wehrtechnische Monatshefte, 65 (Dezember 1968) Heft 12, S. 509-515, Verlag E.S. Mittler und Sohn, Frankf./M.

[3] H.B. Fallmeier: Numerische Beschreibung einiger Zielflugbahnen, Diplomarbeit an der Hochschule der Bundeswehr Hamburg, 1980.

[4] D. Topler: Extrapolation von Zielflugbahnen mit P-, D-, I-, PD- und PID-Korrektur, Studienarbeit an der Hochschule der Bundeswehr Hamburg, 1979.

[5] H.B. Fallmeier: Extrapolation von Zielflugbahnen mit DI- und PI-Korrektur, Studienarbeit an der Hochschule der Bundeswehr Hamburg, 1979.

[6] K. Nixdorff und E. Schmitt: Zur Verwendung des Digitalrechners in der Vorhaltrechnung, Wehrtechnische Monatshefte 65 (November 1968) Heft 11, S. 466-468, Verlag E.S. Mittler und Sohn, Frankfurt/M.

[7] S. Kempfle: Hochschule der Bundeswehr Hamburg, mündliche Mitteilung an den Verfasser, 1979.

[8] P. Schackmar: Spline-Interpolation von Schußtafelwerten, Diplomarbeit an der Hochschule der Bundeswehr Hamburg, 1980.

Prof. Dr.-Ing. Kurt Nixdorff
Hochschule der Bundeswehr Hamburg
Fachbereich Maschinenbau/Mathematik
Holstenhofweg 85
2000 Hamburg 70

SOME REMARKS ON THE COMPUTATION OF $\sqrt{X}$.

Poul Wulff Pedersen.

For a as an approximation to $\sqrt{A}$, the Newton formula aa := $\frac{1}{2}$(a + A/a) defines the (improved) value aa as the mean value of two approximations a and A/a. This is first examined using simple geometry. Next some approximations are obtained by replacing $\frac{1}{2}$ by a factor p. For example aa := p(a + A/a) and aa := p·a + $\frac{1}{2}$·(A/a) with p = p(a,A) slightly less than $\frac{1}{2}$. From this we are led to some 'formula generating formulas', where iteration is used to obtain new iteration formulas. Next some division-free formulas for $\sqrt{A}$ are examined. Finally it is shown how the same iteration technique as for square roots can be used to obtain approximations for functions like sin, using an operator (integral) equation instead of the algebraic equation used for $\sqrt{A}$.

1. The geometry of Newton's formula.

Let A denote the number for which we want the square root, and let a denote an approximation to $\sqrt{A}$:

$$a = \text{APPROX}[\sqrt{A}]$$

Then A/a is also an approximation

$$\frac{A}{a} \simeq \frac{A}{\sqrt{A}} = \sqrt{A}$$

and since

$$a \cdot \frac{A}{a} = \sqrt{A} \cdot \sqrt{A} = A$$

it follows, that if a is too small, i.e. < $\sqrt{A}$, then A/a will be too big, and vice versa. Thus - for the moment for lack of better - we could use the mean

$$\frac{1}{2}(a + \frac{A}{a}) \qquad (1)$$

as a new (and we may hope better) approximation. (1) is of course the usual Heron-Newton value

$$\text{NEW}[a] := \frac{1}{2}\left[a + \frac{A}{a}\right] \qquad (2)$$

The relation between the approximations a, A/a and the value $\sqrt{A}$ is shown below. In particular it is seen where each is best, using as standard notation

$$E[x] = \text{ERROR}[x] := \sqrt{A} - x$$

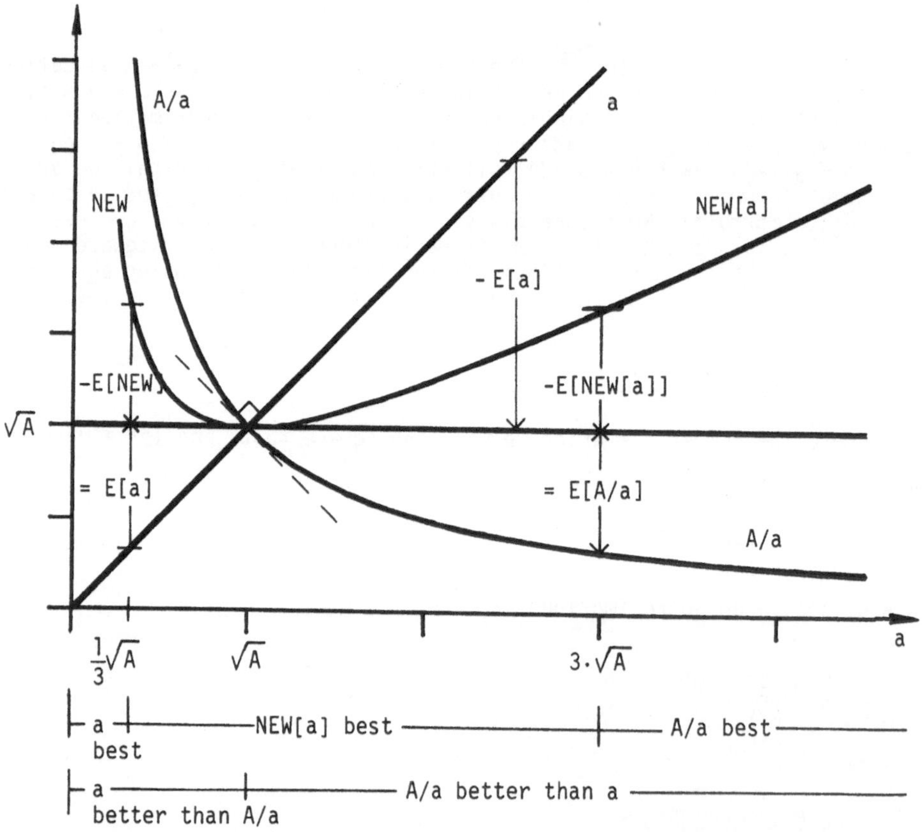

NEW[a] always $\geq \sqrt{A}$, $=$ only for $a = \sqrt{A}$.

2. The factor 1/2 and the error E[NEW].

The 'mean value' argument used here - and it is of course only one of many - can easily be extended. Let E denote the error of an approximation a

$$E := \sqrt{A} - a$$

Then

$$a = \sqrt{A} - E \tag{3}$$

and

$$\frac{A}{a} = \frac{A}{\sqrt{A} - E} = \frac{A - E^2 + E^2}{\sqrt{A} - E} = \sqrt{A} + E + \frac{E^2}{\sqrt{A} - E} \tag{4}$$

From (3), (4)

$$\sqrt{A} = a + E\,, \qquad \sqrt{A} = \frac{A}{a} - E - \frac{E^2}{\sqrt{A} - E} \qquad (5),(6)$$

If we take the mean value, the first order errors E will cancel

$$\sqrt{A} = \frac{1}{2}(a + \frac{A}{a}) - \frac{E^2}{2(\sqrt{A} - E)}$$

and at the same time we get the error E[NEW[a]] of the Newton value

$$E[NEW(a)] = \frac{-E^2}{2(\sqrt{A} - E)}\,, \qquad ER[NEW(a)] = \frac{-(ER)^2}{2(1 - ER)} \qquad (7),(8)$$

ER is the relative error.

3. Replacing 1/2 in Newton's formula by p.

Since the Newton value $aa := \frac{1}{2}(a + A/a)$ will - for $a \neq \sqrt{A}$ - always be too large, we could try to replace 1/2 by a factor p = p(a,A) determined so that we get a better approximation

Let first

$$aa_p := p(a + \frac{A}{a}) \tag{9}$$

The optimal value p* is obtained for aa_p being equal to $\sqrt{A}$. Thus

$$p^* = \frac{\sqrt{A}}{a + \frac{A}{a}}$$

However, since we want p* in terms of quantities which we can readily compute, let

$$e := A - a^2 , \qquad r := \frac{e}{a^2} \tag{11}$$

Here e 'measures' the deviation (error) of a from the correct value - although in a nonlinear way compared with the error E. r is a corresponding relative value.

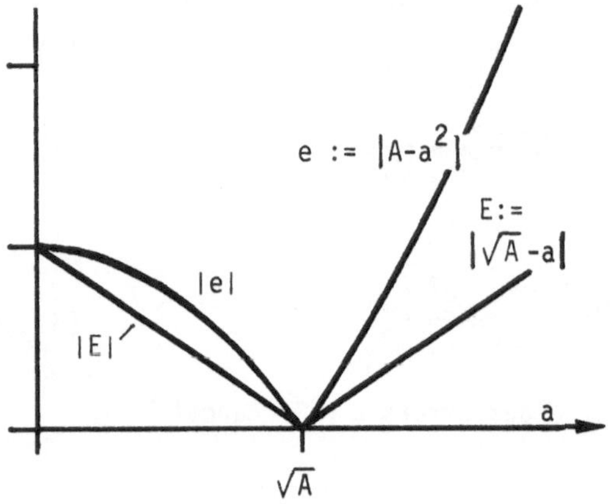

Using (11) we get

$$p^* = \frac{\sqrt{1 + r}}{2 + r} \tag{12}$$

$$= \frac{1}{2}(1 + r)^{\frac{1}{2}}(1 + \tfrac{1}{2}r)^{-1}$$

which gives

$$p^* = \frac{1}{2} - \frac{1}{16} r^2 + \frac{1}{16} r^3 - \frac{13}{256} r^4 + \frac{5}{128} r^5 - \frac{61}{2048} r^6 + - \dots \tag{13}$$

$p = \frac{1}{2}$ gives the Newton value.

Replacing 1/2 by p, part 2.

As an alternative we could give the term a (which for $a > \sqrt{A}$ is the inferior one) a lower weigth factor

$$aa_p := p \cdot a + \frac{1}{2} \frac{A}{a} \tag{14}$$

The optimal value p* is again obtained for $aa_p = \sqrt{A}$, and we get

$$p^* = \frac{\sqrt{A}}{a} - \frac{A}{2 \cdot a^2} = \sqrt{1 + r} - \frac{A}{2 \cdot a^2} = \sqrt{1 + r} - \frac{1}{2}(1 + r) \tag{15}$$

giving

$$p^* = \frac{1}{2} - \frac{1}{8} r^2 + \frac{1}{16} r^3 - \frac{5}{128} r^4 + \frac{7}{256} r^5 - \frac{21}{1024} r^6 + - \dots \tag{16}$$

As before, 1/2 gives the Newton value. The formulas could be used with enough terms to give the accuracy needed in one step, but they could of course also be used iteratively. The p-formulas (13), (16) will only function efficiently for reasonably small values of r, i.e. for reasonably good start approximations. Otherwise p* must be computed in a different way - and in the end it is of course only a question of approximating the p* function, and the Taylor polynomials (13), (16) may not be the best tool for that.

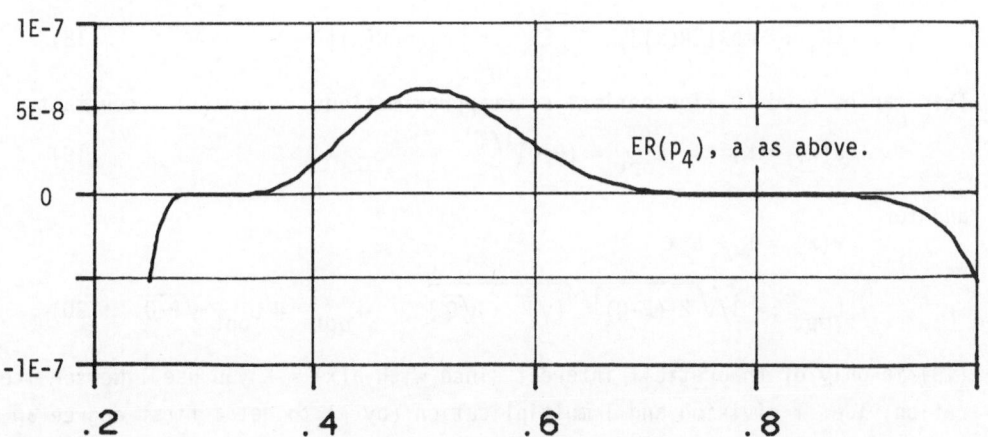

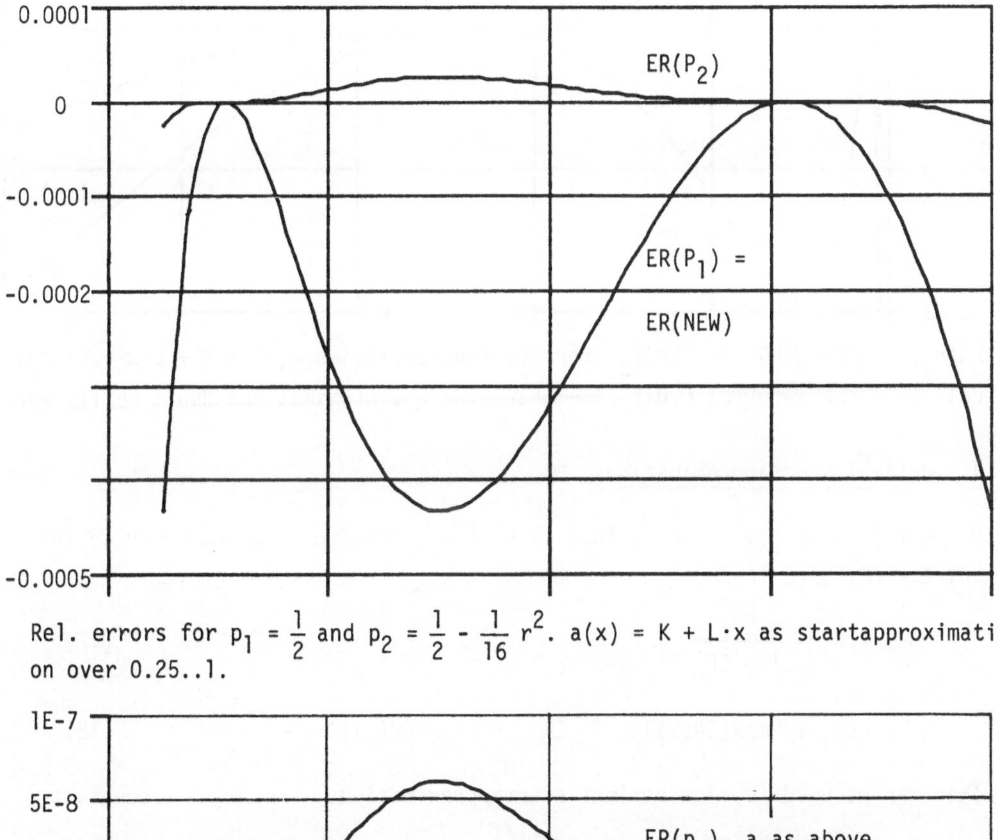

Rel. errors for $p_1 = \frac{1}{2}$ and $p_2 = \frac{1}{2} - \frac{1}{16} r^2$. a(x) = K + L·x as startapproximation over 0.25..1.

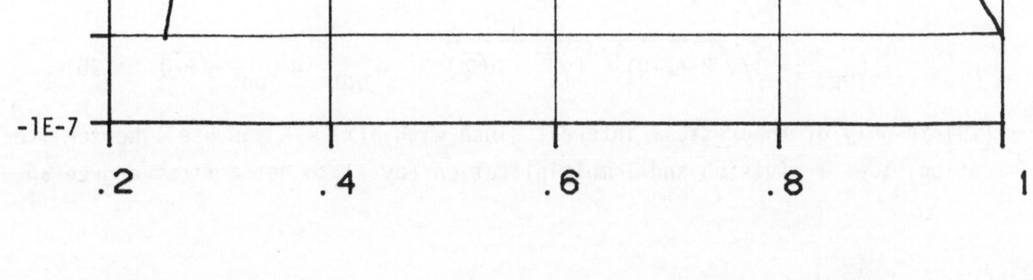

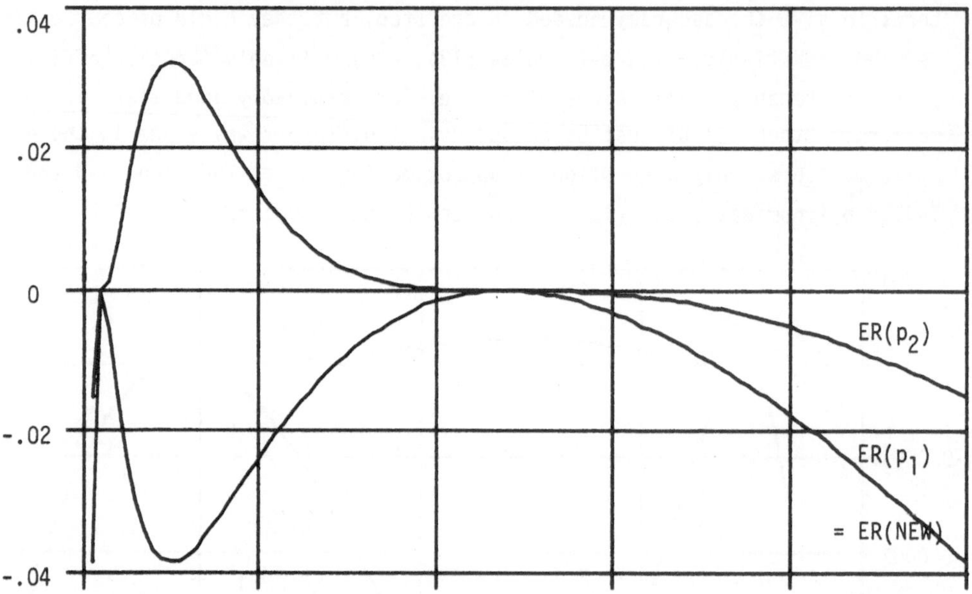

$ER(p_1)$ = ER(NEWTON) and $ER(p_2)$ for startapproximation $a(x) = K + L\cdot x$ (optimal, rel) over the interval 0.01...1 - where the p-approximations don't really work.

4. Optimal startapproximations.

Using (8) it is easily seen, that if an approximation is minmax over an interval I = [P, Q] , then

$$ER_+ = \frac{ER_-}{1 + ER_-} \tag{17}$$

$$ER_+ := \max[ER(x)], \qquad ER_- := \left|\min ER(x)\right| \tag{18}$$

This can be used to find optimal startapproximations. For a(x) = constant

$$a(x) = K, \qquad K_{opt} = (P\cdot Q)^{1/4} \tag{19}$$

and for

$$a(x) = K + L\cdot x$$

$$L_{opt} := 1\Big/\sqrt{2\cdot(P\cdot Q)^{1/4}[\sqrt{P} + \sqrt{Q}]}\ , \qquad K_{opt} := L_{opt}\cdot\sqrt{P\cdot Q} \tag{20}$$

(19) is only of theoretical interest since with a(x) = K you use 1 Newton iteration, i.e. 1 division and 1 multiplication (by ½) to get a first degree ap-

proximation K + L·x - which is not as good as (20). For the more complicated startapproximation types no explicit expressions have been found.

5. A p-factor over an interval.

The error formula (8) shows that a Newton iteration aa = aa(x) will have $ER_+ = 0$. According to (17) it will therefore not be an optimal startapproximation for further Newton iteration. However, by multiplying by a factor p

$$aa_p := p \cdot aa$$

we can 'shift' aa down so that (17) holds for aa_p. The optimal p-value can be found using (17)

$$p^* := 1/\sqrt{1 + ER_-}\,, \qquad ER_- := \left| \min ER[aa(x)] \right| \tag{21}$$

For a given startapproximation a = a(x), the p* values can be found in advance. It also turns out that the exact p* values can - without too great a deterioration - be replaced by SA-numbers, i.e. numbers with which a multiplication can be carried out as (simpler) Shifts and Addition/subtractions. At each step the error wil roughly be halved, so with the square in (8) the cumulative effect will soon be considerable. Compared with ordinary Newton we get an improvement by factors 2, 8, 128, 32768, .. after 1, 2, 3, 4 .. steps, and besides the error left will be (roughly) evenly distributed around 0 instead of being onesided as in the Newton case.

6. A 'formula generating formula'.

The formula

$$\sqrt{A} = \frac{A + a\sqrt{A}}{a + \sqrt{A}} \tag{22}$$

is a (trivial) identity. But if we replace $\sqrt{A}$ on the left- and right hand side by $APPROX_{n+1}$, $APPROX_n$ respectively, writing

$$APPROX_{n+1} := \frac{A + a \cdot APPROX_n}{a + APPROX_n} \tag{23}$$

we get a 'formula producing formula' in the following sense : Let a denote an approximation to $\sqrt{A}$, and let $APPROX_n = APPROX_n(a,A)$ denote an expression which in terms of a and A gives an approximation to $\sqrt{A}$. Then $APPROX_{n+1}$ will be a better - by order 1 - approximation. If for example $APPROX_1 := a$, then $APPROX_2$ will be the Newton approximation, $APPROX_3$ will be the Halley third order formula (which Traub nominated as the most often rediscovered formula !) $APPROX_4$ will be a wellknown 4'th order formula and

$$APPROX_5 = a \cdot \frac{5\,A^2 + 10\,a^2 \cdot A + a^4}{A^2 + 10 \cdot a^2 \cdot A + 5\,a^4} \tag{24}$$

Due to the special coefficients this formula might be of interest for a decimal computer - in this or some other form, for example

$$a^* := a + a \cdot \frac{8 \cdot (A + aa) \cdot (A - aa)}{2 \cdot AA + 10 \cdot aa \cdot (aa + 2A)} \qquad xx := x \cdot x \tag{24a}$$

and with $2\cdot$ and $8\cdot$ obtained by addition ($8 = 10 - 2$). The relative error is

$$\simeq \frac{1}{16}\,(ER)^5 , \qquad ER := ER[a]$$

- and we use only one division.

7. Some division-free formulas.

The AEGP (for Anderson, Earle, Goldschmidt, Powers) algorithm for $\sqrt{A}$ is (essentially) the following :

START : $x_0 := A, \quad d_0 := A, \quad r_0 := 1 + \frac{1}{2}(1 - d_0)$

ITERATION : $x_{n+1} := x_n \cdot r_n$

$d_{n+1} := d_n \cdot (r_n)^2$

$r_{n+1} := 1 + \frac{1}{2}(1 - d_n)$

RESULT : $x_n \rightarrow \sqrt{A}. \quad (d_n \rightarrow 1, \quad r_n \rightarrow 1).$

The r_n part is hardware induced (complement, binary shift and add 1), so for the mathematics we can eliminate r. Furthermore to get faster convergense we can notice that the essential part of the start approximation is that

$$\frac{{x_0}^2}{d_0} = A .$$

Modified AEGP algorithm for $\sqrt{A}$:

START : $k := \text{APPROX}[1/\sqrt{A}]$

$x_0 := k \cdot A$, $d_0 := k^2 \cdot A$

ITERATION : $x_{n+1} := x_n \cdot (\frac{3}{2} - \frac{1}{2} d_n)$

$d_{n+1} := d_n \cdot (\frac{3}{2} - \frac{1}{2} d_n)^2$

This can be further generalised by replacing $\frac{3}{2} - \frac{1}{2} d_n$ by any other function $f = f(d_n)$ so that $d_n \to 1$:

START : $k := \text{APPROX}[1/\sqrt{A}]$, $x_0 := k \cdot A$, $d_0 := k^2 \cdot A$ (25)

ITERATION : $x_{n+1} := x_n \cdot f(d_n)$, $d_{n+1} := d_n \cdot [f(d_n)]^2$, $n = 0,1,2,..$ (26)

with

f chosen so that $d_n \to 1$

Proof :

$$d_{n+1} = \frac{d_{n+1}}{d_n} \frac{d_n}{d_{n-1}} \cdots \frac{d_1}{d_0} \cdot d_0$$

$$= f_n^2 \cdot f_{n-1}^2 \cdots f_0^2 \cdot d_0 \qquad f_i := f(d_i) \tag{27}$$

Since $d_n \to 1$ it follows that

$$f_n \cdot f_{n-1} \cdots f_0 \to 1/\sqrt{A}$$

Similarly

$$x_{n+1} = \frac{x_{n+1}}{x_n} \cdot \frac{x_n}{x_{n-1}} \cdots \frac{x_1}{x_0} \cdot x_0 = f_n \cdot f_{n-1} \cdots f_0 \cdot x_0 \tag{28}$$

which

$$\to x_0/\sqrt{d_0} = \sqrt{A}$$

i.e.

$$x_n \to \sqrt{A} .$$

In particular it is easily seen that for $f(d) = \frac{3}{2} - \frac{1}{2} d$, d_n will $\to 1$. A higher order iteration is obtained for

$$f(d) := \frac{1}{8}(15 - 10 \cdot d + 3 \cdot d^2) \tag{29}$$

Error analysis :

According to the defining equations

$$x_{n+1} = (f_n \cdot f_{n-1} \cdots f_0) \cdot x_0$$

$$= \sqrt{d_{n+1}} \cdot \frac{x_0}{\sqrt{d_0}}$$

$$= \sqrt{d_{n+1}} \cdot \sqrt{A}$$

so that

$$x_n = \sqrt{A} \cdot \sqrt{d_n} \qquad (30)$$

Since $x_n \to \sqrt{A}$ and $d_n \to 1$, the errors are given by

$$E[x_n] = \sqrt{A} - x_n , \qquad E[d_n] = 1 - d_n$$

From (30)

$$\sqrt{A} - x_n = \sqrt{A}(1 - \sqrt{d_n})$$

i.e.

$$E[x_n] = \sqrt{A}(1 - \sqrt{1 - E[d_n]}) \qquad (31)$$

and for small values of $E[d_n]$

$$E[x_n] \simeq \frac{1}{2} \sqrt{A} \cdot E[d_n] , \qquad \text{and} \qquad ER[x_n] \simeq \frac{1}{2} \cdot ER[d_n] \qquad (32)$$

8. Using iteration to obtain approximations for transcendental functions.

Using Green's function, sin(x) can be obtained as a solution to an equation

$$y = F[y]$$

with F as an integral operator. The equation $y_{n+1} := F[y_n]$ can then be used to obtain better (polynomial) approximations.

Poul Wulff Pedersen
Dept. of mathematics,
D.T.H., The technical university of Denmark,
DK-2800 Lyngby (Copenhagen), Denmark.

APPROXIMATE MATRIX INVERSION BY AGGREGATION

Olga Pokorná - Irena Prágerová

The so called aggregation method has been suggested in [1] for inverting block matrices which may be split into the sum of two matrices, one of them being invertible and the other one being "Block-wise constant", i.e. having in each of its blocks all elements of the same value. The method consists in replacing the inversion of a large block matrix by the inversion of a matrix of a lower dimension. It was studied in [3] from the numerical point of view and several numerical examples have been performed. Some results concerning the applicability of this method - in particular to certain class of sparse Leontjev's matrices - are presented in this paper.

Let $A = (a_{ij})$ be a given $n \times n$ matrix. Its block structure may be desribed by the decomposition of the index set $N = \{1, 2, \ldots, n\}$ into r disjoint subsets N_k, $N = \bigcup_{k=1}^{r} N_k$, N_k having n_k elements, $\sum_{k=1}^{r} n_k = n$. Let $n_k \geq 2$ at least for one k.

In this way, A is divided into $r \times r$ blocks $(r < n)$, block A_{pq} consisting of elements a_{ij} with $i \in N_p$ and $j \in N_q$. For simplicity, we suppose that

$N_1 = \{1,2,\ldots,n_1\}$, $N_2 = \{n_1 + 1, \ldots, n_1 + n_2\}$, $\ldots$, $N_r = \{(\sum_{j=1}^{r-1} n_j) + 1, \ldots, n\}$. (This may be reached by eventual rows and columus permutations).

DEFINITION 1

A is called block - wise constant if (for each $k, l = 1,2,\ldots, r$) $a_{ij} = a_{kl}$ for $i \in N_k$ and $j \in N_l$. A is called nearly block-wise constant if for each $k, l = 1,2,\ldots,r$ there exists a constant c_{kl} such that the difference $a_{ij} - c_{kl}$ is small compared with a_{ij} for all $i \in N_k$ and $j \in N_l$.

Let us denote by $\mathcal{L}$ (resp. $\tilde{\mathcal{L}}$) the class of all block-wise constant (resp. nearly block-wise constant) n x n matrices (with the fixed r x r block-wise constant) n x n matrices (with the fixed r x r block structure).

In describing the aggregation method, the concepts given by the following definition are used :

DEFINITION 2

Summation of A is the r x r matrix $s(A) = (s_{kl})$ with

$$s_{kl} = \sum_{p \in N_k, q \in N_l} a_{pq} , \quad k, l = 1, 2, \ldots, r . \tag{1}$$

Condensation of A is the r x r matrix $c(A) = (c_{kl})$ with

$$c_{kl} = \frac{1}{n_k n_l} s_{kl} , \quad k, l = 1, 2, \ldots, r \tag{2}$$

(s_{kl} as in (1)).

The theoretical background of the aggregation method ([1]) may be formulated as follows :

Let H and S be two n x n matrices, H invertible an $S \in \mathcal{L}$. Then H - S is invertible iff

(3) $I - s(H^{-1})\, c\,(S)$

is invertible. In this case,

(4) $(H-S)^{-1} = H^{-1} + H^{-1}\, B\, H^{-1}$,

where $B \in \mathcal{L}$ is such that

(5) $c(B) = c(S)\, [I - s(H^{-1})\, c\,(S)]^{-1}$.

(as usually, I denotes the identity matrix of the corresponding dimension).

Thus, if A = H-S is such that H^{-1} is easily computable (or even known) and $S \in \mathcal{L}$, we may compute the inverse of the n x n matrix A by computing the inverse of the (smaller) r x r matrix (3) and using (4) with (5).

The numerical experiments confirm that the accuracy of A^{-1} computed in this way is the same as when A is inverted directly.

In investigating the applicability of the aggregation mwthod to matrices of the form A = H - S with $S \in \tilde{\mathcal{L}}$, we use the concept of the perturbation of a matrix :

DEFINITION 3

Perturbation of A is the n x n matrix z(A) = (z_{ij}) with

(6) $z_{ij} = a_{ij} - c_{kl}$ for $i \in N_k$ and $j \in N_l$, $k,l = 1,2,\dots,r$

(c_{kl} as in (2)).

If the matrix A to be inverted is of the form $A = H - S$ where H is easily invertible and $S \in \tilde{\mathcal{L}}$, we may try to approximate A^{-1} by the inverse of the matrix $\tilde{A} = H - \tilde{S}$ with $\tilde{S} = S - z(S)$. As $\tilde{S} \in \mathcal{L}$, the aggregation method may be used to invert $\tilde{A}$.

In our experiments we have turned our attention to the approximate inversion of sparse Leontjev's matrices (i.e. sparse matrices $M = I - A$, where all $a_{ij} \geq 0$ and $\sum_{j=1}^{n} a_{ij} < 1$, $i = 1,2,\ldots,n$) of the form just mentioned. Such matrices occur in mathematical models of some economical problems and these problems motivated the idea to transfer the economical concept of aggregation into mathematical analog of their matrix representation.

Let $M = I - A$ be a given Leontjev's matrix with a fixed r x r block structure. Since Leontjev matrix is invertible, M^{-1} exists.

If $A \in \tilde{\mathcal{L}}$, we approximate M by $\tilde{M} = M + z(A) = I - (A - z(A))$. It may be shown that $\tilde{M}$ is Leontjev's, too ([3]). Hence $\tilde{M}^{-1}$ exists.

The splitting of $\tilde{M}$ into the form $\tilde{M} = H - S$, $S \in \mathcal{L}$ may be chosen in several ways. The first way we have considered is to take $H = I$ and $S = A - z(A) \in \mathcal{L}$. In this case, the formula (4) becoms very simple:

$$(I - S)^{-1} = I + B \text{ with } B \in \mathcal{L} \text{ and}$$

$$c(B) = c(S)\left[I - s(I)\, c(S)\right]^{-1},$$

where $s(I) = \mathrm{diag}(n_1, n_2, \ldots, n_r)$.

The second way of splitting M we have used is to take $H = \mathrm{diag}(\tilde{M}_1, \tilde{M}_2, \ldots, \tilde{M}_r)$, where $\tilde{M}_k$ is the k - th diagonal block of $\tilde{M}$, $k = 1,2,\ldots r$. All M_k^{-1} exists and $H^{-1} = \mathrm{diag}(\tilde{M}_1, \tilde{M}_2^{-1}, \ldots, \tilde{M}_r^{-1})$.

In this case $S = H - \tilde{M}$.

The following way also was considered. First, the given matrix M is split, $M = D - \hat{A}$, where $D = \text{diag}\,(M_1, M_2, \ldots, M_r)$, M_k being the k - th diagonal block of M. The inverse of M is then approximated by the inverse of $\tilde{M} = D - (\hat{A} - z(\hat{A}))$. We put $H = D$ and $S = \hat{A} - z(\hat{A})$. This way might provide a better accuracy but unfortunately the existence of $\tilde{M}^{-1}$ is not guarrantied in general.

As every sparse matrix may be transformed by rows and columus permutations to a block-triangular form, we performed our numerical experiments on Leontjev's block-triangular matrices. For testing the guality of the approximation, the given matrices also were inverted directly by a process suggested in [2] for block-triangular Leontjev matrices.

For various n, the process was started with $M = I - A$, $A \in \mathcal{L}$ and then A was modified by adding some matrix P to A. The matrices P were taken with zero row sums. Such matrices P represented the perturbations of the new matrices.

The accuracy of the computed approximation $\tilde{M}^{-1}$ of M^{-1} has been tested by the residual matrix $R = I - M\tilde{M}^{-1}$ or by the error matrix $M^{-1} - \tilde{M}^{-1}$, where M^{-1} had been computed directly as suggested in [3].

For the simplest choice $H = I$ and $S = A - z(A)$ the following theoretical error estimates ([3]) hold :

$$\| I - M\tilde{M}^{-1} \| \leq \| z(A) \| \, (1 + \| K \|), \text{ where}$$

$I + K = (I - S)^{-1}$, and

$\| M^{-1} - \tilde{M}^{-1} \| < \| z(A) \| \quad (1 + 2\|K\|) + o(q^2)$, where $q = \| (I - S)^{-1} z(A) \|$ and it is supposed $q < 1$.

As may be seen in the table, the order of the largest error in the residual matrix does not exceed the order of the largest elements of the perturbation matrix for the computed numerical examples.

dim. of A (n)	dim of diag. blocks (n_k)	highes order of elem. in A	highest order of elem. in P	number of nouzero elem. of P in % of n
50	12,6,2, 10,13,7	10^{-4}–10^{-2}	10^{-4}–10^{-3}	3 %
100	5,7,18,7 4,2,7,3, 10,9,11,10	10^{-4}–10^{-3}	10^{-4}–10^{-3}	3 %
100	5,7,18,7 4,2,7,3, 10,9,11,10	10^{-4}–10^{-3}	10^{-4}–10^{-3}	15%

highes order of elem. in R	highes order of elem. in S
10^{-3}	10^{-3}
10^{-3}	10^{-3}
10^{-3}	10^{-3}

Some investigations were perfomed concerning the influence of the block structure of M onto the accuracy of the approximation of M^{-1}. The accuracy may be improved by making the blockstructure denser if the corresponding perturbance matrix in the denser structure has smaller elements in comparison with the orginal structure.

The numerical experiments have shown that the aggregation method may be applied to approximate inversion of matrices of some classes, especially of certain type of sparse Leontjev materices. The accuracy of the approximation in the investigated cases was of the same order as in results obtained by direct inversion. Timeconsumming can be reduced by the aggregation method for large sparse matrices of the appropriate class.

References :

1 Fiedler, M. - Pták V. : On Aggregation in Matrix Theory and Its Application to Numerical Inverting of Large Matrices, Bulletin de l'acad. Polon. des Sc., Série des sc. math., astr. et phys., vol. XI, No 12 (1963) 757 - 759

2 Nováková, M. : On the Use Of Some Properties Of Leontjev's Matrices In Their Inversion, Aplikace matematiky, Svazek 15, č. 2, (1970)

3 Prágerová, I. : Použití agregace při inversi

matic, Praha (1980), (Diploma work)

RNDr. Olga Pokorná,CSc., RNDr. Irena Prágerová
KNM FMF UK
Malostranské nám. 25
118 00 PRAHA 1
Czechoslovakia

FLOPPY VS. FUSSY APPROXIMATION

Rudolf Scherer and Karl Zeller

The approximation of multivariate functions is in general a difficult task, especially if one wants to obtain optimal solutions (the characterizing H-sets are mostly hard to find). But there are many cases where satisfactory approximations can be established with little expenditure. Thus the practical complexity of such problems is low. We propose to apply especially simple methods for computing "floppy" approximations: Green Mathematics for developing areas. One point is to employ rather coarse grids, in connection with error estimates for the meshes. Another idea is to improve initial approximations (e.g. determined by interpolation) by correcting polynomials of the single peak type.

1. Introduction

The title is perhaps provocative: Are mathematicians either sloppy or pedantic? If a reader should think so we revoke and retract to Gauß, who was against excessively accurate computations. In the computer age his point of view might be challenged. But the fact remains that many computations are surplus and can distract from developing better methods and theories. We mention some examples connected with approximation and optimization (aiming at large scale problems).

The Simplex method dominated practical linear programming, despite certain drawbacks. But gradually other methods (like Triplex, Hungarian, Ellipsoid) enter into the picture. They can be viewed as stable procedures achieving an optimal solution within reasonable time, but also as tools working out acceptable solutions within short time. These remarks lead over to numerical approximation (primarily in the case: linear subspace, discrete variables).

Assuming the Haar-Chebyshev condition (hence considering mainly the univariate case) it is not too difficult to find an optimal solution by applying an ascent algorithm (Remez; cf. the books by Meinardus [6], Collatz and Krabs [2] and Watson [10]). The corresponding multivariate problems are much more complicated in general (cf. Watson [10] pp. 72, 80): The number of grid-points can be very large, the characteristic H-set might be hard to recognize (degeneracies), instabilities have to be watched. Hence there is some sense in looking for robust rules of thumb giving handy answers without trouble and finesse: Green mathematics for developing areas. Such rules might then lead by gradual refinement to new theories.

2. Approximation

We approximate functions F (real-valued, continuous, with period 2π) by trigonometric polynomials Q of degree at most m (the distance given by the max-norm). We use

$$D(x) := \frac{2}{2m+1} \frac{\sin \frac{2m+1}{2} x}{2 \sin \frac{1}{2} x}$$

(the Dirichlet polynomial), the equidistant nodes

$$x_j := \frac{2j\pi}{2m+1} \qquad (j=-m,\dots,m)$$

and the matching basic polynomials for interpolation:

$$D_j(x) := D(x-x_j) \qquad (j=-m,\dots,m)$$

(vanishing at the nodes x_k $(k \neq j)$ and with value 1 at x_j).

The considerations extend to the bivariate (multivariate) case by means of tensor products (cf. Haußmann and Pottinger [5] and Scherer and Zeller [7],[8]).

Our interpolation process yields an almost best approximation (loss described by a logarithmic factor). An actual check of the accuracy is easy to conduct, if we assume that F is (replaced by) a polynomial and thus employ a grid G of r equidistant nodes w_k , requiring

$$\deg Q \leq m , \quad \deg F \leq n , \quad \# G = r \quad (m \leq n < \frac{r}{2}) ,$$

$$w_k = w_1 + \frac{(k-1)2\pi}{r} \quad (k=1,\dots,r) , \qquad s := \frac{\pi}{r} .$$

Then we have (for the Chebyshev norm, putting $P := F-Q$)

$$\|P\| \leq C \max_k |P(w_k)| , \text{ where } \quad C := \{\cos \frac{n\pi}{r} \}^{-1}$$

(cf. for example Scherer and Zeller [7], where also the bivariate case is treated).

Our next goal is to refine the last estimate. Then we proceed to improve the approximation at critical points.

3. Error Estimates

Our estimates go back to M. Riesz and S. Bernstein. They depend on the fact that the polynomial cos nx has certain extremal properties (longest alternation, steepest descent, maximal derivative).

LEMMA 1. If $\deg P \leqslant n$ and $s^* := ns$ and

$P(w)=\|P\|$ for some $w \in [w_1, w_2]$,

$P(w_1)=1$, $P(w_2)=1-d$ $(0 \leqslant d \leqslant 2\sin^2 s^*)$,

then

$$\|P\|^2 \leqslant \frac{4\sin^2 s^* - 4d\sin^2 s^* + d^2}{4\sin^2 s^* \cos^2 s^*}$$

holds.

The right hand side decreases from $\cos^{-2} s^*$ to 1 (stationary point). For larger d the norm $\|P\|$ is not attained in our interval. The main part of the proof consists in constructing an (extremal) $P(x) = a\cos nx + b\sin nx$ which satisfies the interpolatory equations and then computing $\|P\|^2 = a^2 + b^2$ (by matrix calculus).

LEMMA 2. If $\deg P \leqslant n$ and

$\|P''\| \leqslant 2a$, $P(w_1)=1$, $P(w_2)=1-d$ $(d \geqslant 0)$,

then

$P(x) \leqslant 1 + K$ $(w_1 \leqslant x \leqslant w_2)$,

where

$K := a(s-b)^2$ with $b := \frac{d}{4as}$ (for $0 \leqslant d \leqslant 4as^2$)

and $K=0$ for larger d .

The value K decreases from as^2 to 0 (stationary point). The proof uses an interpolating quadratic function Q with $Q''=-2a$

A bound 2a follows from the Markov-Bernstein inequality and Lemma 1. Further estimates, also from below, can be obtained by using more points and derivatives.

4. Correctors

The preceding estimates show that for many purposes coarse grids suffice. This is especially important for multivariate problems. If we want to improve an approximation on such a grid then coarse methods might be the proper tools. We keep in mind that besides the error on the grid we have to watch the error estimates for the intervals. So we propose a descent algorithm using correcting polynomials with one main hump (peak).

First we consider polynomials of the Zolotarev type: equioscillation with one larger hump. These can be obtained from squeezed Chebyshev polynomials:

$$Z_\rho(x) := T(-1+\rho(u+1)) \ , \ u = \cos x \quad (\rho > 1) \ .$$

Of course we modify Z_ρ by scale factors and shifts. The parameter ρ adjusts the corrector to the shape of the error curve (near an extremum point). The secondary peaks of Z_ρ near 0 can be useful; often they adjust to the error curve. Then other correctors with more suitable wave forms might give better results, e.g.

$$D_o(x+t) \ , \ D_1 + D_2 \ , \quad -D_o + D_1 + D_2 - D_3$$

up to the extremal polynomial D^* giving the norm of the interpolation operator. One basic idea is to find better interpolation nodes. Another idea subsists on changing the values on the nodes or on the grid (regarding storage requirements).

In the bivariate case we obtain correctors as products of two polynomials, primarily in the tensor case (rectangle), but with easy generalizations.

5. Numerical Experience

We have tried these methods on many examples, taken from the literature (e.g. Watson [9]) or designed for our needs. The examples included univariate and bivariate functions with different smoothness properties (the grid had to be chosen correspondingly).

Our initial approximation was determined by interpolation. Thus the first error function often showed a rather regular behavior, with one main peak (or symmetric peaks) and damped oscillation. In such cases the correctors of type D (see above) were quite effective, especially the extremal polynomial D^*. But in general it is easier to apply correctors of type Z. We can choose ρ such that again $\max_x Z_\rho(x) = \|I_m\|$ (the norm of the interpolation operator), decreasing ρ in critical situations. Finally there are singular cases with violently oscillating error; this indicates that not much improvement is possible (utilize estimates as above).

It is good to have some lower bound for the approximation degree $E_m(f)$. Bounds can be obtained by the alternation principle or by corresponding linear functionals (based on H-sets). These estimates are be applied to specify the scalar factor to the basic corrector and to stop the process. Often we arrive at the error range indicated by a coarse grid after a few iterations (the number increases moderately with m, n, r). See also the next section.

In many cases it is possible and appropriate to store the values of the (multivariate) function F on a coarse grid. Sometimes reductions to univariate functions are feasible. For the approximation process it suffices in principle to know the values of D on a suitable grid.

6. Remarks

The numerical examples together with theoretical reflections show some benefits, shortcomings and desirable improvements of the plain or floppy approach (in contrast with subtle or fussy methods). In many cases one achieves a rather good approximation (a presolution) with little effort and without stability problems. Thus the practical complexity of such problems is low. But once we arrive at an error function with many extremum points it might be better to treat all these points simultaneously (as in common descent methods). For high accuracy one will prefer adaptive grids, ascent methods and equations for the points of the H-set (cf. Watson [10]pp. 72, 80 ; Blatt [1]).

Looking at the grid points, we have to solve an overdetermined system of linear equations in the Chebyshev sense (alternatively with modifications in view of the meshes). The use of the basic correction polynomials can be described as a transformation strengthening the main diagonals. The choice of the scalar factor offers problems as in known relaxation methods (over, under, optimum?).

Now we turn to multivariate approximation (mainly in the tensor product case). The tensor grids and the corresponding corrector polynomials utilize the information contained in the topology of the problem (it is simpler than one might suspect). Linear programming methods in general disregard this aspect; they can further have some trouble with instabilities and near-singularity of basis matrices (cf. Watson [10] p. 72). The use of fixed grids presents some opportunities to determine H-sets in advance or by non-numerical calculations.

We hope that the floppy vs. fussy debate leads to a better handling and understanding of some numerical problems.

References

[1] Blatt, H.P.: A general Remes algorithm in real or complex normed linear spaces. In: Handscomb, D.C. (ed.): Multivariate Approximation. London-New York-San Francisco, Academic Press 1978.

[2] Collatz, L., Krabs, W.: Approximationstheorie. Stuttgart, Teubner 1973.

[3] Defert, Ph., Thiran, J.P.: Chebyshev approximation by multivariate polynomials. Publications du département mathématiques, Facultés universitaires de Namur. Report 80/10 (1980).

[4] Ehlich, H., Zeller, K.: Numerische Abschätzung von Polynomen. Z. Angew. Math. Mech. 45, T20-T22 (1965).

[5] Haußmann, W., Pottinger, P.: On the construction and convergence of multivariate interpolation operators. J. Approx. Theory 19, 205-221 (1977).

[6] Meinardus, G.: Approximation von Funktionen und ihre numerische Behandlung. Berlin-Göttingen-Heidelberg-New York, Springer 1964.

[7] Scherer, R., Zeller, K.: Two-dimensional grids for polynomials. In: Schempp, W., Zeller, K. (eds.): Multivariate Approximation Theory. ISNM 51, pp. 346-352, Basel-Boston-Stuttgart, Birkhäuser 1979.

[8] Scherer, R., Zeller, K.: Gestufte Approximation in zwei Variablen. In: Collatz, L., Meinardus, G., Werner, H. (eds.): Numerische Methoden der Approximationstheorie, Band 5. ISNM 52, pp. 282-288, Basel-Boston-Stuttgart, Birkhäuser 1980.

[9] Watson, G.A.: A multiple exchange algorithm for multivariate Chebyshev approximation. SIAM J. Numer. Anal. 12, 46-52 (1975).

[10] Watson, G.A.: Approximation Theory and Numerical Methods. Chichester-New York-Brisbane-Toronto, John Wiley and Sons 1980.

Rudolf Scherer
Karl Zeller
Mathematisches Institut
Universität Tübingen
Auf der Morgenstelle 10
D-7400 Tübingen

EIN PARAMETERIDENTIFIZIERUNGSPROBLEM AUS DER PULSRADIOLYSE

Rita Schmidt

For the investigation of very fast chemical reactions often pulsradiolytic methods are applied. The underlying mathematical model is based on reaction kinetics, where the rate constants of the reactions are to be determined. The usual treatment of this problem using a nonlinear least squares fit leads to uncertainties concerning the quality of the solution. In the following controlltheoretic methods are applied to give characterizing equations which can be used to verify the solution. This requires the treatment of systems of coupled nonlinear boundary value problems.

1. EINFÜHRUNG

Die vorliegenden Untersuchungen wurden für die Herstellung eines Routineprogramms durchgeführt, das die Auswertung von Daten gestattet, die durch Experimente im Bereich Strahlenchemie des Hahn-Meitner-Instituts anfallen. Das Ziel dieser Experimente ist die Aufklärung schnell ablaufender chemischer Prozesse, die mit pulsradiolytischen Methoden untersucht werden.

Das Bild 1 möge diesen Sachverhalt veranschaulichen: Eine wäßrige Lösung befindet sich im chemischen Gleichgewicht. Es ist bekannt, welche Substanzen in der Lösung enthalten sind. Ebenso ist das Verhältnis der Konzentrationen dieser Substanzen in der

Lösung bekannt. Durch einen Strahlpuls hochenergetischer Elektronen wird das Gleichgewicht gestört.

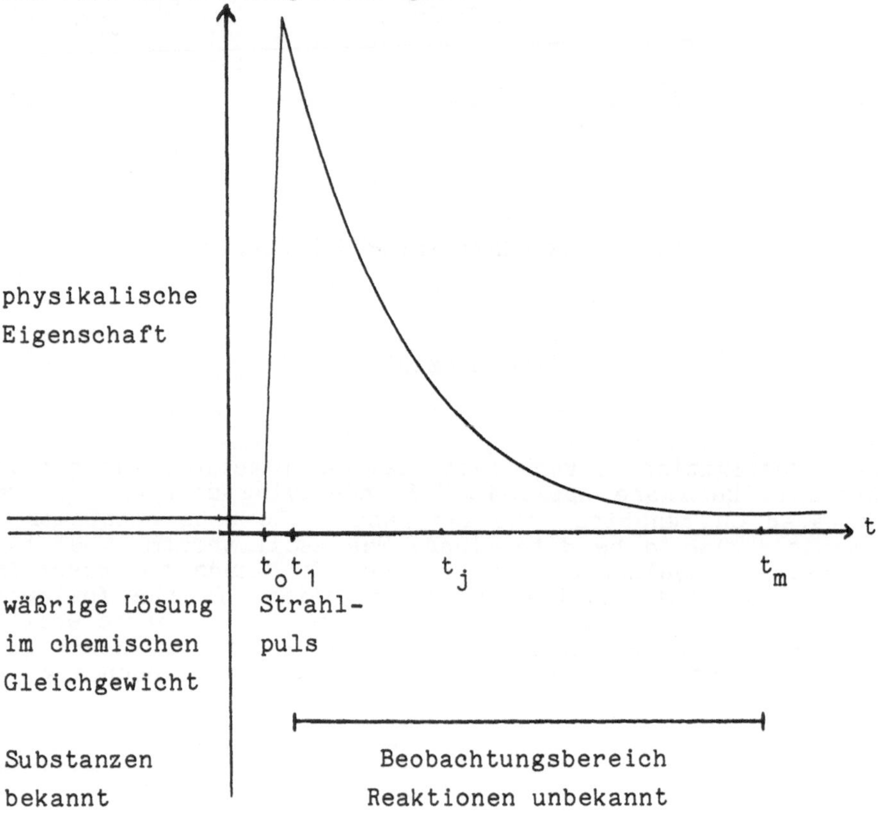

Bild 1: Experimentablauf

Es finden chemische Reaktionen statt, die es aufzuklären gilt. Da die Reaktionen sehr schnell ablaufen - die Strahlzeit dauert 10^{-12} bis 10^{-9} Sekunden, die Lösung ist in 10^{-2} Sekunden wieder im Gleichgewicht - kommen chemische Analysen zur Untersuchung nicht in Betracht. Es ist jedoch möglich, die durch die Reaktionen hervorgerufenen Änderungen physikalischer Eigenschaften der Lösung in diesen kurzen Zeiten zu messen. Insbesondere sind das Messungen der optischen Absorption und der elektrischen Leitfähigkeit. Es werden an einer Lösung i.a. mehrere Versuche mit unterschiedlichen Strahlintensitäten oder unter Beobachtung anderer physikalischer Eigenschaften durchgeführt, die natürlich alle durch ein einziges Reaktionsschema beschrieben werden sollen. Ein typischer Datensatz für eine solche Versuchsausführung ist im Bild 2 dargestellt.

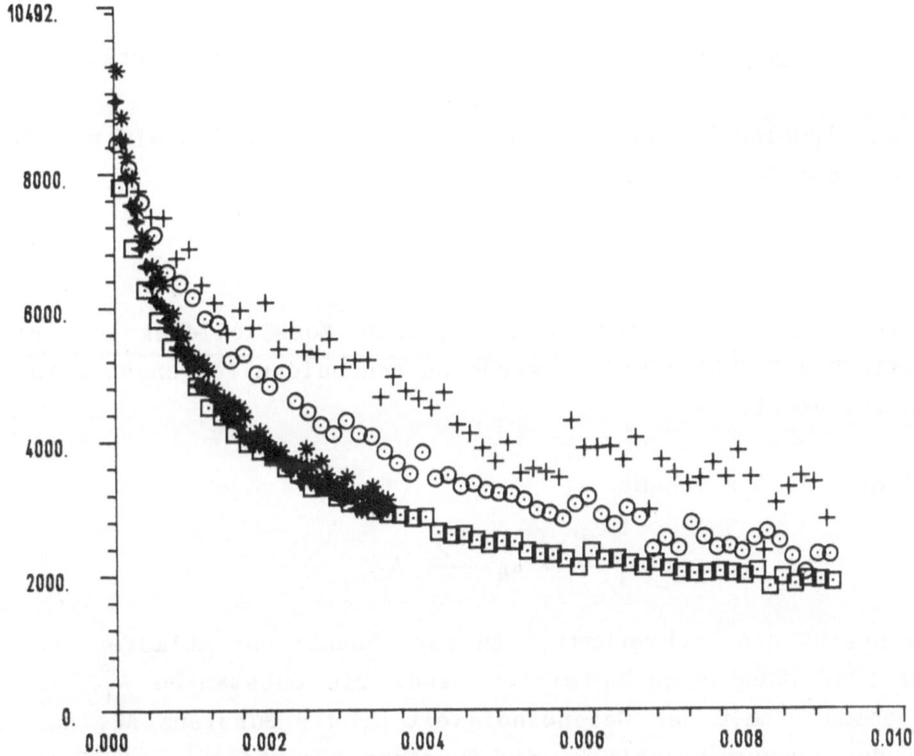

Bild 2: 5 Meßreihen zur Identifizierung einer chemischen Reaktion

Die Plotsymbole ✱,+,⊙,⊡,♦ kennzeichnen die zu einer Messung gehörigen Datenpunkte. Es wurden fünf Versuche an einer Reaktionslösung durchgeführt. Sie sind durch die folgenden Daten charakterisiert:

	PS	$10^7 c_A$	m_σ	$10^{-3} z_M$	$10^4 t_1$	$10^2 t_m$	$10^4 \Delta t$
1	*	2.7761	54	9.538	0.44	0.3436	0.64
2	+	0.5189	56	8.490	1.55	0.8955	1.60
3	⊙	1.1000	57	8.451	0.35	0.8995	1.60
4	⊡	2.7761	57	7.801	0.70	0.9030	1.60
5	♦	2.7273	54	9.083	0.28	0.3420	0.64

Tabelle 1: Die Daten des Identifizierungsproblems

Das Problem besteht nun darin, für die Meßdaten ein passendes Reaktionsmodell zu finden.

2. MODELLBILDUNG

Der Mechanismus der mathematischen Modellbildung aus einem chemischen Reaktionsmodell werde an den beiden folgenden Beispielen erläutert.

Beispiel 1: Das Schema

$$A_1 + A_2 + A_3 \xrightarrow{k_1} A_4 \xrightarrow{k_2} A_5$$

beschreibt den Sachverhalt, daß zwei Reaktionen ablaufen, an denen fünf Substanzen beteiligt sind. Die Substanzen A_1, A_2 und A_3 erzeugen mit der Geschwindigkeit k_1 die Substanz A_4, aus der mit der Geschwindigkeit k_2 die Substanz A_5 entsteht. Die Reaktionen können nur dann ablaufen, wenn die zugehörigen Anfangskonzentrationen a_1, a_2 und a_3 von Null verschieden sind.

Das mathematische Modell beschreibt die zeitlichen Änderungen der Konzentrationen der Substanzen. Diese sind der Reaktionsgeschwindigkeit und den Konzentrationen der an der Reaktion beteiligten Stoffe direkt proportional, abnehmend für reagierende und zunehmend für entstehende Stoffe.

Für das obige Schema ergibt sich somit das System

$$\frac{dA_1}{dt} = - k_1A_1A_2A_3$$

$$\frac{dA_2}{dt} = - k_1A_1A_2A_3$$

$$\frac{dA_3}{dt} = - k_1A_1A_2A_3$$

$$\frac{dA_4}{dt} = k_1A_1A_2A_3 - k_2A_4$$

$$\frac{dA_5}{dt} = k_2A_4 \quad .$$

Beispiel 2: Der Reaktion

$$A_1 + A_2 \xrightarrow{k_1} A_2$$

wird das Gleichungssystem

$$\frac{dA_1}{dt} = - 2k_1{A_1}^2$$

$$\frac{dA_2}{dt} = k_1{A_1}^2$$

zugeordnet, da zwei Mole der Substanz A_1 benötigt werden, um ein Mol von A_2 zu bilden.

Chemische Reaktionen lassen sich also durch Systeme gewöhnlicher Differentialgleichungen beschreiben. Diese haben die Form

$$(1) \quad \begin{aligned} & y' = A(y)k, \; y(t)\in\mathbb{R}^n, \; t\in\mathbb{R}, \; k\in\mathbb{R}^r, \; A\colon \mathbb{R}^n \rightarrow \mathbb{R}^{n\times r}, \\ & y(t_0) = ca, \; c\in\mathbb{R}, \; a\in\mathbb{R}^n. \end{aligned}$$

Darin ist r die Zahl der ablaufenden Reaktionen, n die Zahl der beteiligten Substanzen.

In den hier vorliegenden Anwendungen wird c als Normierungsfaktor für die Pulsintensität verwendet. Der Vektor a gibt die feste Verteilung der Anfangskonzentrationen an. Die Experimentdaten (s. Bild 2) lassen sich in der Form

$$D := \{D^{(\sigma)},\ \sigma = 1,\dots,s\} \quad \text{mit}$$

$$(2) \qquad D^{(\sigma)} := \{c_A^{(\sigma)}, (t_j^{(\sigma)}, z_j^{(\sigma)}),\ j = 1,\dots,m_\sigma,$$

$$c_A^{(\sigma)} \in \mathbb{R},\ (t_j^{(\sigma)}, z_j^{(\sigma)}) \in \mathbb{R}\times\mathbb{R}\ \}$$

schreiben. $c_A^{(\sigma)}$ ist ein Normierungsfaktor.

Der Zusammenhang zwischen den Experimentdaten $(t,z(t))$ und der Lösung $y(t)$ des Prozeßmodells

$$y' = A(y)k,\ y(t_o) = ca$$

wird durch das Experimentmodell

$$z(t) = \alpha^T y(t),\ \alpha \in \mathbb{R}^n,\ z(t) \in \mathbb{R},\ t \in \mathbb{R}$$

hergestellt. Der Vektor α gibt an, in welchen Anteilen die einzelnen Substanzen erfaßt werden, wenn eine physikalische Eigenschaft der Lösung gemessen wird.

Obwohl es möglich wäre, aus der Kenntnis der Anfangskonzentrationen in der ungestörten Lösung eine Folge von Reaktionsmodellen automatisch aufzubauen, um aus den Meßdaten ein passendes Modell zu identifizieren, ist es für Routinearbeiten sinnvoller, dem Experimentator die Wahl eines speziellen Reaktionsschemas zu überlassen. Die Identifizierung beschränkt sich dann auf die Bestimmung der Reaktionskonstanten.

Da sich bei der vorliegenden Anwendung die Strahlintensitäten nicht genau genug bestimmen lassen, werden sie auch in die Identifizierung einbezogen. Hingegen werden die Vektoren $a \in \mathbb{R}^n$ und

$\alpha \in \mathbb{R}^n$ als exakt vorgebbar behandelt. Unter diesen Voraussetzungen erlauben die Daten die Formulierung des folgenden nichtlinearen Ausgleichsproblems:

Minimiere

$$f(y,k,c) := \frac{1}{2} \sum_{\sigma=1}^{s} \sum_{j=1}^{m_\sigma} (z_j^{(\sigma)} - \frac{\alpha^T}{c_A^{(\sigma)}} y^{(\sigma)}(t_j^{(\sigma)}))^2$$

(3) unter den Nebenbedingungen

$$y^{(\sigma)}(t_o) = c^{(\sigma)} a$$

$$y^{(\sigma)\prime}(t) = A(y^{(\sigma)}(t))k \text{ für } t \in [t_o, t_{m_\sigma}^{(\sigma)}] \text{ und } \sigma = 1, \ldots, s.$$

Die direkte Behandlung dieses Problems mit ableitungsfreien Methoden führt zu Unsicherheiten über die Güte der erzielten Lösung. Deshalb werden durch Anwendung kontrolltheoretischer Methoden Bedingungsgleichungen für eine Optimallösung hergeleitet, deren Erfülltsein numerisch überprüft werden kann.

3. KONTROLLTHEORETISCHE BEHANDLUNG

Der Einfachheit halber werde zuerst $s = 1$ gewählt. Der Produktraum hat dann die Form

$$X := Y \times \mathbb{R}^n \times \mathbb{R} \quad \text{mit} \quad Y := C_n^1 [t_o, t_m] .$$

Ein Punkt $x = (y,k,c)$ heißt zulässig, wenn er die Gleichung (1) erfüllt. Ein Punkt $x_o = (y_o,k_o,c_o)$ heißt optimal, wenn in (3)

$$f(y_o,k_o,c_o) \le f(y,k,c)$$

für alle zulässigen $x \in X$ gilt.
Es liegen die beiden Abbildungen vor:

$f: X \to \mathbb{R}$, f ist durch (3) definiert
$g: X \to Y$ mit

$$g(y,k,c) := y(\bullet) - ca - \int_{t_0}^{\bullet} A(y(\tau))kd\tau$$

für alle $(y,k,c)\in X$.

Das ist die Prozeßgleichung in integrierter Form.

Das Minimierungsproblem lautet:
Minimiere $f(y,k,c)$ unter der Nebenbedingung $g(y,k,c) = \Theta$, $\Theta\in Y$.

Zur Gewinnung charakterisierender Gleichungen wird nun das Maximumprinzip herangezogen (s. [1] und [2]). Da Ungleichungsnebenbedingungen fehlen - die Forderungen $k\geq\Theta, c\geq\Theta$ sind durch die Wahl passender Näherungen stets erfüllt - hat es die folgende Form:

Wenn eine Optimallösung $x_0 =(y_0,k_0,c_0)$ und die Fréchet-Ableitungen

$$f'_{x_0}(x) = f'_{(y_0,k_0,c_0)}(y,k,c) \quad \text{und}$$

$$g'_{x_0}(x) = g'_{(y_0,k_0,c_0)}(y,k,c)$$

in diesem Punkt x_0 existieren, dann gibt es
- eine nicht negative reelle Zahl l_0 und
- ein reelles lineares Funktional l auf Y, die

nicht beide identisch Null sind, so daß die Funktionalrelation

$$\langle l, g'_{(y_0,k_0,c_0)}(y,k,c)\rangle = l_0 f'_{(y_0,k_0,c_0)}(y,k,c)$$

für alle $y\in Y$, $k\in R^r$, $c\in R$ und $t\in[t_0,t_m]$ erfüllt ist.

Im konkreten Problem existieren die Fréchet-Ableitungen. Da g'_{x_0} surjektiv ist, darf o.B.d.A. $l_0 = 1$ gewählt werden. Somit hat die Funktionalrelation die Form

$$\langle l, y(\cdot) - \int_{t_0}^{\cdot} h_0(\tau)y(\tau)d - \int_{t_0}^{\cdot} A(y_0(\tau))kd\tau - ca\rangle$$

(4)

$$= \sum_{j=1}^{m} \chi_0(t_j)\left(-\frac{\alpha^T}{c_A}y(t_j)\right) \quad \text{für alle } y\in Y,\ k\in R^r,\ c\in R$$

mit

$$h_o(t) := (A(y_o(t))k_o)_y^T \text{ , } h_o(t) \text{ ist (nxn)-Matrix,}$$

$$\chi_o(t) := z(t) - \frac{\alpha^T}{c_A} y_o(t) \text{ , } \chi_o(t) \in \mathbb{R}.$$

Aus dieser Funktionalrelation werden nun durch Spezialisierung

- ein System adjungierter Gleichungen
- eine Kontrollbedingung und
- eine Orthogonalitätsbedingung

gewonnen, die zusammen mit der Nebenbedingung aus (3) ein nichtlineares Randwertproblem bilden, das im Optimalpunkt

$$x_o = (y_o, k_o, c_o)$$

notwendigerweise erfüllt sein muß. Dadurch kann die Bestimmung einer Optimallösung durch die numerisch zuverlässigere Methode einer Gleichungsauflösung vorgenommen werden.

Zur Herleitung dieser Gleichungen ist es notwendig, eine Darstellung des linearen Funktionals l für ein Element $v \in Y$ zu finden. Da die rechte Seite der Funktionalrelation (4) von k und c unabhängig ist, genügt es, diese für das spezielle Element

$$x = (y, \Theta, 0)$$

zu betrachten. Dann gilt

$$(5) \quad < l, y(\cdot) - \int_{t_o}^{\cdot} h_o(\tau) y(\tau) d\tau > = \sum_{j=1}^{m} \chi_o(t_j) (- \frac{\alpha^T}{c_A} y(t_j)).$$

Aus

$$v(t) := y(t) - \int_{t_o}^{t} h_o(\tau) y(\tau) d\tau$$

ergibt sich durch Auflösung nach y

$$y(t) = v(t) + \Phi(t) \int_{t_o}^{t} \Phi^{-1}(\tau) h_o(\tau) v(\tau) d\tau \text{ ,}$$

worin Φ das System der Fundamentallösungen der homogenen Gleichung $y' = h_o y$ ist. Wird in (5) y durch v ersetzt, so hat das lineare Funktional l die folgende Darstellung

$$(6) \qquad < l,v(\cdot)> = \sum_{j=1}^{m} \{\chi_o(t_j)(-\frac{\alpha^T}{c_A}v(t_j)) + \int_{t_o}^{t_j} r_j(t)v(t)dt\}.$$

Darin ist

$$r_j(t) := -\chi_o(t_j)\frac{\alpha^T}{c_A}\Phi(t_j)\Phi^{-1}(t)h_o(t),$$

$$r_j(t)^T \in \mathbb{R}^n, \quad t \in [t_o,t_j].$$

Für $\quad v(t) = \int_{t_o}^{t} w(\tau)d\tau$

ergibt sich aus (6) die Gleichung

$$(7) \qquad <l, \int_{t_o} w(\tau)d\tau >= -\sum_{j=1}^{m} \int_{t_o}^{t_j} p_j(t)w(t)dt$$

mit

$$(8) \qquad p_j(t) := \chi_o(t_j)\frac{\alpha^T}{c_A} - \int_{t}^{t_j} r_j(\tau)d\tau \text{ für } t\in[t_o,t_j],\ j=1,\ldots,m.$$

Die Gleichungen (6) und (7) lassen sich nun verwenden, um aus der Funktionalrelation (4) durch Spezialisierung der Komponenten von $x \in X$ die charakterisierenden Gleichungen für eine Optimallösung zu gewinnen. Dazu werden jeweils zwei der drei Komponenten Null gesetzt, wodurch die Funktionalrelation (4) eine spezielle Gestalt erhält. Auf diese werden die Gleichungen (6) und (7) angewandt, und es ergeben sich Bedingungen, die im Optimalpunkt notwendigerweise erfüllt sein müssen.

Sei $x = (y,\Theta,0)$. Aus (4) folgt (5). Die Anwendung von (6) und

(7) auf die linke Seite von (5) ergibt

$$(9) \qquad \langle 1, y(\cdot) - \int_{t_o}^{\cdot} h_o(\tau)y(\tau)d\tau \rangle = \sum_{j=1}^{m} \{\chi_o(t_j)\left(-\frac{\alpha^T}{c_A}y(t_j)\right)$$

$$+ \int_{t_o}^{t_j} r_j(t)y(t)dt + \int_{t_o}^{t_j} p_j(t)h_o(t)y(t)dt\} .$$

Aus dem Vergleich der rechten Seiten von (5) und(9) folgt

$$\sum_{j=1}^{m} \int_{t_o}^{t_j} \{p_j'(t) + p_j(t)h_o(t)\}y(t)dt = 0$$

mit $p_j' = r_j$ aus (8).

Diese Gleichung muß für alle $y \in Y$ erfüllt sein.
Eine Umordnung des obigen Ausdrucks derart, daß die Quadraturen intervallweise ausgeführt werden, führt zu der Beziehung

$$\sum_{l=1}^{m} \int_{t_{l-1}}^{t_l} \sum_{j=l}^{m} (p_j'(t) + p_j(t)h_o(t))y(t)dt = 0.$$

Diese kann durch Einführung der Sprungfunktion

$$(10) \qquad p(t) := \sum_{j=l}^{m} p_j(t) \ , \quad t \in (t_{l-1}, t_l], \ l=1,\dots,m,$$

in der Form

$$\sum_{l=1}^{m} \int_{t_{l-1}}^{t_l} (p'(t) + p(t)h_o(t))y(t)dt = 0$$

oder kürzer

$$\int_{t_o}^{t_m} (p'(t) + p(t)h_o(t))y(t)dt = 0$$

geschrieben werden. Da diese Gleichung für alle $y \in Y$ gelten soll, ist notwendig, daß

$$p'(t) + p(t)h_o(t) \equiv \Theta$$

ist für alle $t \in [t_o, t_m]$.
Aus (10) und (8) folgen die Sprungrelationen

$$p(t_j^-) - p(t_j^+) = p_j(t_j) = \chi_o(t_j)\frac{\alpha^T}{c_A}$$

für $j = m-1,\ldots,1$ und die Anfangsbedingung

$$p(t_m) = p_m(t_m) = \chi_o(t_m)\frac{\alpha^T}{c_A}$$

für den rechten Randpunkt $t = t_m$.

In den Problemdaten (s. (4)) formuliert, liegt also das adjungierte System vor:

$$p^{T\prime}(t) = -(A(y_o(t))k_o)_y p^T(t) \quad \text{für } t_m \geq t \geq t_o$$

mit

$$p^T(t_m) = (z(t_m) - \frac{\alpha^T}{c_A}y_o(t_m))\frac{\alpha}{c_A}$$

und

$$p^T(t_j^-) = p^T(t_j^+) + (z(t_j) - \frac{\alpha^T}{c_A}y_o(t_j))\frac{\alpha}{c_A}$$

für $j = m-1,\ldots,1$.

Dieses homogene lineare Differentialgleichungssystem hat nur dann eine nichttriviale Lösung, wenn mindestens einer der Fehlerterme

$$\chi_o(t_j) = z(t_j) - \frac{\alpha^T}{c_A}y_o(t_j), \quad j \in \{1,\ldots,m\}$$

von Null verschieden ist. Die Kondition des Problems ist umso schlechter, je besser die Daten durch die Lösung

$$z(t) = \frac{\alpha^T}{c_A} y_o(t)$$

approximiert werden. Liegen fehlerfreie Daten vor, so können auf diese Weise keine charakterisierenden Gleichungen gewonnen werden.

Die Kontrollbedingung ergibt sich für $x = (\Theta, k, 0)$ aus (4) und (7) zu

$$<1,- \int_{t_o} A(y_o(\tau))kd\tau> = \sum_{j=1}^{m} \int_{t_o}^{t_j} p_j(t)A(y_o(t))kdt = 0$$

bzw. nach Umordnung und unter Verwendung von (10) zu

$$\int_{t_o}^{t_m} p(t)A(y_o(t))kdt = 0 \text{ für alle } k \in \mathbb{R}^r .$$

Wird insbesondere $k = e_\rho$ für $\rho = 1,\ldots,r$ gewählt, so folgt das nichtlineare Gleichungssystem

$$\int_{t_o}^{t_m} p(t)A(y_o(t))dt = \Theta^T \text{ mit } \Theta \in \mathbb{R}^r .$$

Die Orthogonalitätsbedingung folgt für $x = (\Theta, \Theta, c)$ aus (4), (6), (8) und (10): Es muß

$$<1,- ca> = \sum_{j=1}^{m} p_j(t_o)ca = p(t_o)ca = 0$$

sein für alle $c \in \mathbb{R}$, d.h. aber

$$p(t_o)a = 0 .$$

Bevor die nichtlineare Randwertaufgabe endgültig formuliert werden kann, ist die Erweiterung auf beliebig viele Datenreihen für die Beschreibung eines chemischen Prozesses vorzunehmen. Sei s die Anzahl der Datenreihen, die alle unterschiedlich lang sein dürfen (s. Bild 2). Es muß dann der Produktraum

$$X := Y^{(1)} \times \ldots \times Y^{(s)} \times \mathbb{R}^r \times \mathbb{R}^s \ , \quad Y^{(\sigma)} := C_n^1[t_o, t_{m_\sigma}^{(\sigma)}]$$

für $\sigma = 1,\ldots,s$ zugrundegelegt werden.
Durch den Vektor $k \in \mathbb{R}^r$ werden die an sich unabhängigen Probleme gekoppelt. Es ist $c^T = (c_1,\ldots,c_s)$ der Vektor der Anfangsintensitäten.
Die Abbildungen f und g haben je s Komponenten:

$$f: X \to \mathbb{R}$$

mit

$$f := \sum_{\sigma=1}^{s} f^{(\sigma)},$$

$$f^{(\sigma)}(y^{(\sigma)},k,c^{(\sigma)}) := \frac{1}{2} \sum_{j=1}^{m_\sigma} (z_j^{(\sigma)} - \frac{\alpha^T}{c_A^{(\sigma)}} y^{(\sigma)}(t_j^{(\sigma)}))^2,$$

$$g : X \to Y \ , \ Y = Y^{(1)} \times \ldots \times Y^{(s)},$$

mit

$$g(y,k,c) := (y^{(1)}(\cdot) - c^{(1)}a - \int_{t_o}^{\cdot} A(y^{(1)}(\tau))k\,d\tau, \ldots,$$

$$y^{(s)}(\cdot) - c^{(s)}a - \int_{t_o}^{\cdot} A(y^{(s)}(\tau))k\,d\tau)$$

für alle $(y^{(1)},\ldots,y^{(s)},k,c) \in X$.
Die Fréchet-Ableitungen von f und g sind komponentenweise zu bilden. In der Funktionalrelation

$$< l,g' > = l_o f'$$

hat l die Form $l = (l^{(1)},\dots,l^{(s)})$. Mit $l_0 = 1$ ist

$$\langle l,g'\rangle = \sum_{\sigma=1}^{s} \langle l^{(\sigma)},g^{(\sigma)\prime}\rangle = \sum_{\sigma=1}^{s} f^{(\sigma)\prime}.$$

Die Herleitung der notwendigen Bedingungen geschieht wie im Fall einer Datenreihe durch Spezialisierung des Elementes $x \in X$. Dazu werden in $x = (y^{(1)},\dots,y^{(s)},k,c)$ jeweils alle Komponenten bis auf eine Null gesetzt.
Für jedes $y^{(\sigma)} \neq \Theta$, $\sigma = 1,\dots,s$, ergibt sich ein System adjungierter Gleichungen der Form

$$p^{(\sigma)\prime}(t) = - p^{(\sigma)}(t)(A(y_0^{(\sigma)}(t))k_0)_y{}^T \text{ für } t \in [t_0, t_{m_\sigma}^{(\sigma)}]$$

mit

$$p^{(\sigma)}(t_{m_\sigma}^{(\sigma)}) = (z_{m_\sigma}^{(\sigma)} - \frac{\alpha^T}{c_A^{(\sigma)}} y^{(\sigma)}(t_{m_\sigma}^{(\sigma)}))\frac{\alpha^T}{c_A^{(\sigma)}}$$

und

$$p^{(\sigma)}(t_j^{(\sigma)-}) = p^{(\sigma)}(t_j^{(\sigma)+}) + (z_j^{(\sigma)} - \frac{\alpha^T}{c_A^{(\sigma)}} y^{(\sigma)}(t_j^{(\sigma)}))\frac{\alpha^T}{c_A^{(\sigma)}}$$

Für $k \neq \Theta$ ergibt sich die Kontrollbedingung

$$\sum_{\sigma=1}^{s} \int_{t_0}^{t_{m_\sigma}^{(\sigma)}} p^{(\sigma)}(t)A(y_0^{(\sigma)}(t))dt = \Theta^T, \ \Theta \in \mathbb{R}^r.$$

Die Kopplung der Probleme durch den Vektor k kommt dadurch zum Ausdruck, daß die Kontrollbedingungen der Einzelprozesse addiert werden.
Für $c \neq \Theta$ ergeben sich die s unabhängigen Orthogonalitätsbedingungen

$$p^{(\sigma)}(t_o)a = 0 \text{ für } \sigma = 1,\dots,s.$$

Durch die Anwendung kontrolltheoretischer Methoden auf das vorliegende Problem läßt sich nun die folgende nichtlineare Randwertaufgabe zur Charakterisierung einer Optimallösung angeben. Für $\sigma = 1,\dots,s$ sind die folgenden Bedingungen zu erfüllen:

- die Prozeßgleichung

$$y^{(\sigma)\prime}(t) = A(y^{(\sigma)}(t))k \text{ mit } y^{(\sigma)}(t_o) = c^{(\sigma)}a \quad \text{in } t_o \le t \le t_{m_\sigma}^{(\sigma)}$$

- die adjungierte Gleichung

$$p^{(\sigma)\prime}(t) = -\,p^{(\sigma)}(t)(A(y^{(\sigma)}(t))k)_y^T \text{ in } t_{m_\sigma}^{(\sigma)} \ge t \ge t_o$$

mit

$$p^{(\sigma)}(t_m^{(\sigma)}) = (z^{(\sigma)}(t_{m_\sigma}^{(\sigma)}) - \frac{\alpha^T}{c_A^{(\sigma)}}y^{(\sigma)}(t_{m_\sigma}^{(\sigma)}))\frac{\alpha^T}{c_A^{(\sigma)}}$$

und

$$p^{(\sigma)}(t_j^{(\sigma)-}) = p^{(\sigma)}(t_j^{(\sigma)+})$$
$$+ (z^{(\sigma)}(t_j^{(\sigma)}) - \frac{\alpha^T}{c_A^{(\sigma)}}y(t_j^{(\sigma)}))\frac{\alpha^T}{c_A^{(\sigma)}}$$

- die Orthogonalitätsbedingung

$$p^{(\sigma)}(t_o)a = 0$$

sowie

- die Kontrollbedingung

$$\sum_{\sigma=1}^{s} \int_{t_o}^{t_{m_\sigma}^{(\sigma)}} p^{(\sigma)}(t)A(y^{(\sigma)}(t))dt = \Theta^T, \quad \Theta\in\mathbb{R}^r.$$

Diese Randwertaufgabe enthält

- Integralbedingungen für die Lösung
- Unstetigkeiten in der Lösung der adjungierten Gleichung, die

von System zu System wechseln

- in verschiedenen Intervallen definierte Systeme, die über den Vektor k gekoppelt sind.

Deshalb werden keine Standardverfahren zur Lösung nichtlinearer Randwertaufgaben eingesetzt, sondern es wird das folgende Minimierungsproblem gelöst:

Minimiere

$$RES(k,c) := \left\| \sum_{\sigma=1}^{s} \int_{t_o}^{t_m^{(\sigma)}} {}^{\sigma}p^{(\sigma)}(t)A(y^{(\sigma)}(t))dt \right\|^2 + \sum_{\sigma=1}^{s} (p^{(\sigma)}(t_o)a)^2$$

bzgl. $k \in \mathbb{R}^r$ und $c \in \mathbb{R}^s$.

Dabei wird benutzt, daß für festes k und c die n-Vektorfunktionen $y^{(\sigma)}$ und $p^{(\sigma)}$ durch die Differentialgleichungssysteme eindeutig bestimmt sind. Für dieses Zielfunktional ist bekannt, daß es in einem Optimalpunkt den Wert Null hat, wodurch ein numerisch überprüfbares Abbruchkriterium für ein Iterationsverfahren gefunden worden ist.

4. NUMERISCHES BEISPIEL

Als Beispiel sei die Anwendung der Methode auf die im Bild 2 gegebene Datenmenge gezeigt.

In einer wäßrigen Lösung sind die Substanzen A_1 und A_2 mit den Konzentrationen $a_1 = 3.2$, $a_2 = 2.7$ enthalten. Daraus läßt sich primär das Reaktionsschema

$$A_1 + A_2 \xrightarrow{k_1} A_3$$

$$A_1 + A_1 \xrightarrow{k_2} A_4$$

$$A_2 + A_2 \xrightarrow{k_3} A_5$$

aufbauen. Die Matrix des Prozeßmodells hat die Form

$$A(y) = \begin{pmatrix} -y_1y_2 & -2y_1^2 & 0 \\ -y_1y_2 & 0 & -2y_2^2 \\ y_1y_2 & 0 & 0 \\ 0 & y_1^2 & 0 \\ 0 & 0 & y_2^2 \end{pmatrix}.$$

Gemessen wird $z(t) = \alpha_1 y_1(t) + \alpha_4 y_4(t)$ mit $\alpha_1=2812.5$, $\alpha_2=312.5$. Bild 3 zeigt die Daten und die durch das Verfahren ermittelten besten Lösungen $z^{(\sigma)}$ für $\sigma = 1,\ldots,5$.

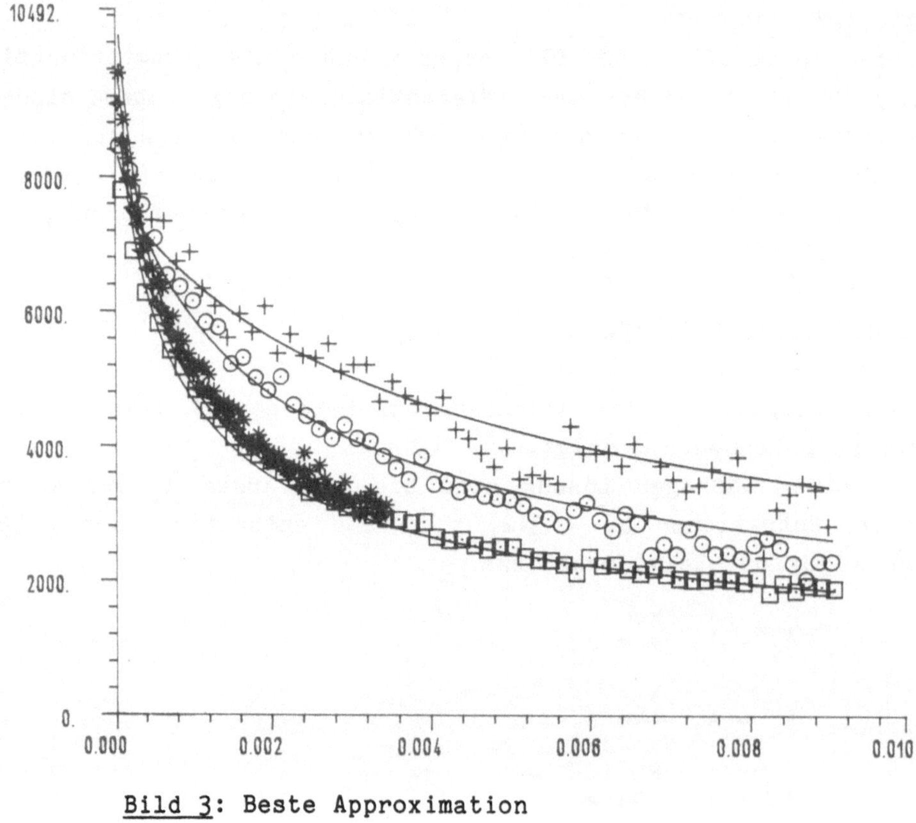

Bild 3: Beste Approximation

L	$10^{-8}k_1$	$10^{-8}k_2$	$10^{-8}k_3$	$10^{-7}F$	RES	FA
1	10.00	10.00	10.00	78.070	$3.27_{10}17$	19
2	17.82	2.59	22.57	1.837	$6.81_{10}14$	16
3	15.47	2.02	13.76	1.623	$1.52_{10}12$	32
4	15.01	1.57	9.17	1.597	$1.51_{10}11$	19
5	14.89	1.38	7.78	1.595	$3.53_{10}07$	23
6	14.88	1.38	7.75	1.595	$2.30_{10}05$	

<u>Tabelle 2</u>: Ergebnisse einer iterierten Minimierung

Tabelle 2 zeigt die Ergebnisse einer iterierten Minimierung. Pro Minimierungslauf L werden die Startwerte der

- Reaktionskonstanten k_1, k_2, k_3
- Fehlerquadratsumme F

$$F := \frac{1}{2} \sum_{\sigma=1}^{s} \sum_{j=1}^{m_\sigma} (z^{(\sigma)}(t_j^{(\sigma)}) - \frac{\alpha^T}{c_A^{(\sigma)}} y^{(\sigma)}(t_j^{(\sigma)}))^2,$$

d.i. der Fehler zwischen den Daten und dem Modell,

- Residuenquadratsumme RES

$$\text{RES} := \left\| \sum_{\sigma=1}^{s} \int_{t_0}^{t_{m_\sigma}^{(\sigma)}} p^{(\sigma)}(t) A(y^{(\sigma)}(t))\,dt \right\|^2$$

$$+ \sum_{\sigma=1}^{s} (p^{(\sigma)}(t_0)a)^2,$$

d.i. der Fehler in der nichtlinearen Randwertaufgabe,

angegeben. Es folgt die Zahl der Funktionsaufrufe FA, die zur Erzielung der in der nächsten Zeile angegebenen Daten erforderlich waren. Es ist zu sehen, daß die Fehlerquadratsumme relativ rasch stabil bleibt, während die Parameter und die Residuenquadratsumme sich noch stark ändern. Die hohe absolute Größe von RES, die ja im Optimalpunkt Null sein soll, resultiert aus notwendig gewesenen Normierungen. Wichtig ist die Abnahme um zwölf Zehnerpotenzen im Laufe der Minimierungen. Der Vektor c ist in der Tabelle nicht enthalten (s. aber [3]). Die Änderungen in den einzelnen Komponenten liegen zwischen 4% und 20%, sie sind also nicht zu vernachlässigen. Weitere Untersuchungen in [3] zeigen, daß auch die Annahme von Folgereaktionen, die ja zwischen den Ausgangs- und den erzeugten Substanzen denkbar sind, zu stabilen Lösungen des Identifizierungsproblems führen. Die dargestellte Methode kann also nicht verwendet werden, um die Güte eines Reaktionsmodells zu beurteilen.

LITERATURVERZEICHNIS

[1] Kirsch, A., Warth, W., Werner, J.: Notwendige Optimalitätsbedingungen und ihre Anwendung. Berlin-Heidelberg-New York, Springer 1978

[2] Lempio, F.: Tangentialmannigfaltigkeiten und infinite Optimierung. Habilitationsschrift, Universität Hamburg 1972

[3] Schmidt, R.: Ein Parameteridentifizierungsproblem aus der Pulsradiolyse. Dissertation, Freie Universität Berlin 1980

Rita Schmidt
Hahn-Meitner-Institut für Kernforschung Berlin GmbH
Bereich Datenverarbeitung und Elektronik
Glienicker Str. 100
D-1000 Berlin 39

KONSTANTEN IM APPROXIMATIONSSATZ VON JACKSON-TIMAN-TELJAKOWSKII

Herbert Friedrich SINWEL *
vorgetragen von Paul Otto RUNCK

In Jackson's theorem on best approximation of continuous functions by algebraic polynomials the factor 1/n can be replaced by the weight function $\sqrt{1-x^2}/n + 1/n^2$ (Timan 1951) or $\sqrt{1-x^2}/n$ (Teljakowskii 1966). These theorems cannot be used in numerical calculation, because they do not give any information about the approximation constants. The purpose of this paper is to establish Teljakowskii's theorem with explicit constants. Furthermore asymptotic best upper bounds in the theorems of Jackson and Timan are presented.

1. PROBLEMSTELLUNG

Die Approximationssätze von Jackson, Timan und Teljakowskii lassen sich folgendermaßen zusammenfassen:

SATZ 1. Sei $\Delta_n(x) := 1/n$ (Jackson)
oder $\Delta_n(x) := \sqrt{1-x^2}/n + 1/n^2$ (Timan)
oder $\Delta_n(x) := \sqrt{1-x^2}/n$ (Teljakowskii)

Dann existiert für jedes $r \in \mathbb{N}$ eine Zahl A_r, sodaß für jedes $f \in C^r[-1,1]$ und jedes $n \in \mathbb{N}$, $n \geq r$, ein Polynom $p \in P_n$ existiert, sodaß für alle $x \in [-1,1]$ gilt:

$$|f(x) - p(x)| \leq A_r(\Delta_n(x))^r w(f^{(r)}, \Delta_n(x))$$

* Diese Arbeit wurde aus Mitteln des österreichischen Fonds zur Förderung der wissenschaftlichen Forschung unterstützt.

Eine etwas schwächere Formulierung verwendet anstatt des Stetigkeitsmoduls den maximalen Betrag der Ableitung. Da die Abschätzungen von Satz 1 außerdem für $n < r$ prinzipiell ungültig sind, liegt es nahe, statt n^{-r} einen Ausdruck zu verwenden, der für $n < r$ divergiert. An anderer Stelle [3,4] wurden folgende Resultate bewiesen, wobei $\|\ \|$ die Supremumsnorm und

$$K_r := \frac{4}{\pi} \sum_{l=0}^{\infty} (-1)^{l(r+1)} (2l+1)^{-r-1}$$

die Konstante von Favard, Achieser und Krein bezeichne.

SATZ 2. Für alle $r,n \in \mathbb{N}$ mit $n \geq r-1$ und alle $f \in C^r[-1,1]$ existiert ein Polynom $p \in P_n$, sodaß

$$\|f - p\| \leq K_r \frac{(n-r+1)!}{(n+1)!} \|f^{(r)}\|$$

SATZ 3. Für alle $r,n \in \mathbb{N}$ mit $n > 2r$ und alle $f \in C^r[-1,1]$ existiert ein Polynom $p \in P_{n-1}$, sodaß für alle $x \in [-1,1]$ gilt

$$|f(x) - p(x)| \leq \frac{K_r}{(n-2)(n-4)\dots(n-2r)} \left(\sqrt{1-x^2} + \frac{2r}{n}|x|\right)^r \|f^{(r)}\|$$

Aus einem Resultat von Bernstein folgt, daß K_r in beiden Fällen die bestmögliche von n unabhängigen Konstanten ist.

Für den Satz von Teljakowskii stellt sich nun folgendes

PROBLEM 1. Für alle $r \in \mathbb{N}$ ist eine Konstante M_r zu bestimmen, sodaß für alle $n \in \mathbb{N}$ mit $n \geq r$ und alle $f \in C^r[-1,1]$ ein Polynom $p \in P_n$ existiert, sodaß für alle $x \in [-1,1]$ gilt:

$$|f(x) - p(x)| \leq M_r \frac{(n-r)!}{n!} \sqrt{1-x^2}^{\,r} \|f^{(r)}\|$$

Da die Menge aller Polynome P dicht liegt in $C^r[-1,1]$ mit der Norm $\|f\|+\|f^{(r)}\|$ braucht das Problem 1 nur für $f \in P$ gelöst werden. Dieselbe Konstante M_r gilt dann für alle $f \in C^r[-1,1]$. Außerdem wird durch die Hintereinanderausführung mit cos das algebraische Problem in ein trigonometrisches überführt, wir erhalten dadurch das zu Problem 1 äquivalente

PROBLEM 2. Für alle $r \in \mathbb{N}$ ist eine Konstante M_r zu bestimmen, sodaß für alle $n \in \mathbb{N}$ mit $n \geq r$ und alle $f \in P$ ein $q \in T_n$ existiert, sodaß für alle $t \in \mathbb{R}$ gilt:

$$|f(\cos t) - q(t)| \leq M_r \frac{(n-r)!}{n!} |\sin t|^r \|f^{(r)}\|$$

Dieses Problem läßt sich leicht in zwei getrennte Aufgaben zerlegen.

PROBLEM 2A. Für jedes $r \in \mathbb{N}$ soll eine Indexmenge Ω_r und zu jedem $f \in P$ eine Familie trigonometrischer Polynome $\{f_\omega : \omega \in \Omega_r\}$ bestimmt werden, sodaß nur endlich viele f_ω nicht verschwinden und weiters gilt:

$$f \circ \cos - \sin^r \sum_\omega f_\omega \in T_r$$

PROBLEM 2B. Es soll eine Konstante M_r bestimmt werden, sodaß für alle $\omega \in \Omega_r$ und $n \in \mathbb{N}$ ein $d_{\omega n} \in \mathbb{R}$ mit folgenden Eigenschaften existiert:

a) Zu jedem $f \in P$ existiert ein $q_{\omega n} \in T_{n-1}$ mit

$$\|f_\omega - q_{\omega n}\| \leq d_{\omega n} \|f^{(r)}\|$$

b) $\sum_\omega d_{\omega n} \leq M_r \frac{(n-1)!}{(n+r-1)!}$

Der Ansatz $q := f \circ \cos - \sin^r \sum_\omega (f_\omega - q_{\omega\, n-r+1})$ zeigt, daß mit 2A und 2B auch das Problem 2 gelöst ist. Um nun für ein beliebiges Polynom f die Funktion $f \circ \cos$ geeignet zerlegen zu können, werden Operatoren J_r benötigt, die $f^{(r)}$ in ein trigonometrisches Polynom überführen, das $\sin^r$ als Faktor enthält und bis auf ein Element T_r mit $f \circ \cos$ übereinstimmt. Bei der Konstruktion der Operatoren J_r hilft

SATZ 4. Für alle $r \in \mathbb{N}$ seien $J_r : P \to T$ so gegeben, daß für alle $f \in P$ gilt:

a) $f \circ \cos - J_1 f' \in T_1$

b) $J_{r+1} f - \frac{1}{r} \cos . J_r f + \frac{1}{r} J_r(\mathrm{id}.f) \in T_{r+1}$

Dann gilt für alle $r \in \mathbb{N}$ und $f \in P$:

$$f \circ \cos - J_r f^{(r)} \in T_r$$

Der Beweis von Satz 4 gelingt mit vollständiger Induktion. Aussage a) bildet die Induktionsbasis, Aussage b) angewandt auf $f^{(r+1)}$ liefert gemeinsam mit der Induktionsvoraussetzung angewandt auf f' und id.f' - r.f den Induktionsschluß.

2. DIE KONSTRUKTION DER INTEGRALOPERATOREN J_r

Es werden zwei Hilfsoperatoren benötigt:

DEFINITION 1. Für $f \in P$ und $k \in \mathbb{N}$ sei

$$\hat{f}_k := \frac{1}{\pi} \int_0^{2\pi} \cos kt . f(\cos t) dt$$

DEFINITION 2. Für jede Folge (a_k), $k \in \mathbb{N}$, sei

$$\Delta a_k := a_k - a_{k+1}$$

BEMERKUNG 1. Aus diesen Definitionen folgt unmittelbar:

a) $f(\cos t) = \sum_{k=1}^{\infty} \hat{f}_k \cos kt + \text{const.} \qquad (f \in P)$

b) Für $k > \deg f$ gilt $\hat{f}_k = 0 \qquad (f \in P)$

c) $\Delta^m a_k = \sum_{j=0}^{m} (-1)^j \binom{m}{j} a_{k+j} \qquad (k, m \in \mathbb{N})$

In dieser Arbeit beziehe sich der Operator Δ definitionsgemäß immer auf den Index k.

DEFINITION 3.

a) Für $k \in \mathbb{N}$ und $p \in \mathbb{N}_o$ sei

$$b_{kp} := \begin{cases} \frac{1}{k+1} & \text{für } p = 0 \\ \frac{1}{k+2p+1} - \frac{1}{k+2p-1} & \text{sonst} \end{cases}$$

b) Für $j = (j_1, \ldots, j_m) \in \mathbb{N}^m$ sei $|j| := \sum_{i=1}^{m} j_i$

Dabei gelte speziell $\mathbb{N}^o := \{\emptyset\}$ und $|\emptyset| := 0$

c) Für $r \in \mathbb{N}$, $m \in \{0, \ldots, r-1\}$ sei $J_{rm} : P \to T$ definiert durch

$$(J_{rm} f)(t) := (\sin t)^r \sum_{k=1}^{\infty} \sum_{p=0}^{\infty} \sum_{j \in \mathbb{N}^m} (\Delta^{r-1+m} b_{k+2|j|-m\ p}) . \hat{f}_{k+2|j|+2p} . \cos(kt + \frac{r}{2}\pi)$$

d) Für $r \in \mathbb{N}$ sei $J_r := \frac{1}{(r-1)!} \sum_{m=0}^{r-1} (-1)^m \binom{r-1}{m} J_{rm}$

Wegen Bemerkung 1b) handelt es sich bei Definition 3c) nur um eine endliche

Summe. In den folgenden Lemmata können die Summen daher beliebig umgeordnet werden.

LEMMA 1. Für alle $f \in P$ gilt $f \circ \cos - J_1 f' \in T_1$.

Beweis: Definition 3 liefert unmittelbar

$$(J_1 f')(t) = \sin t \sum_{k=1}^{\infty} \sum_{p=0}^{\infty} b_{kp} \widehat{f'}_{k+2p} \cos(kt + \tfrac{\pi}{2}) = \ldots =$$

$$= \sum_{k=2}^{\infty} \frac{\cos kt}{2k} (\widehat{f'}_{k-1} - \widehat{f'}_{k+1}) - \frac{1}{4} \widehat{f'}_1 - \frac{1}{6} \widehat{f'}_2 \cos t +$$

$$+ \frac{1}{2} \sum_{k=2}^{\infty} (\frac{1}{k} - \frac{1}{k+2}) \widehat{f'}_{k+1} \cos(k - 2[k/2])t$$

Aus $f(\cos t) = \sum_{k=1}^{\infty} \widehat{f'}_k \cos kt + \text{const.}$ folgt mit der Definition von $\hat{f}_k$ und partieller Integration

$$f(\cos t) = \sum_{k=1}^{\infty} \frac{\cos kt}{2k} (\widehat{f'}_{k-1} - \widehat{f'}_{k+1}) + \text{const.}$$

Daraus folgt die Behauptung des Lemmas.

LEMMA 2. Für alle $r \in \mathbb{N}$, $m \in \mathbb{N}_o$, $m < r$ und $f \in P$ gilt

$$\cos . J_{rm} f - J_{rm}(\text{id}.f) - J_{r+1\ m} f + J_{r+1\ m+1} f \in T_{r+1}$$

Beweis: Mit der Hilfsüberlegung $\widehat{\text{id}.f}_k = (\hat{f}_{k-1} + \hat{f}_{k+1})/2$, der Definition von J_{rm} und einer Verschiebung des Index k ergibt sich, daß sich die Ausdrücke

$$\cos t . J_{rm} f(t) - J_{rm}(\text{id}.f)(t) \qquad \text{und}$$

$$\frac{1}{2}(\sin t)^r \sum_{k=1}^{\infty} \sum_{p=0}^{\infty} \sum_{j \in \mathbb{N}^m} [(\Delta^{r+m} b_{k+2|j|-m-1\ p}) \hat{f}_{k+2|j|+2p-1} -$$

$$- (\Delta^{r+m} b_{k+2|j|-m\ p}) \hat{f}_{k+2|j|+2p+1}] \cos(kt + \frac{r}{2}\pi)$$

nur durch ein trigonometrisches Polynom aus T_{r+1} unterscheiden.

Durch Addition eines geeigneten Polynoms aus T_{r+1} entsteht daraus

$$(\sin t)^{r+1} \sum_{k=1}^{\infty} \sum_{p=0}^{\infty} \sum_{j\in\mathbb{N}^m} \sum_{i=0}^{\infty} [(\Delta^{r+m} b_{k+2|j|+2i-m\ p})\ \hat{f}_{k+2|j|+2i+2p} -$$

$$- (\Delta^{r+m} b_{k+2|j|+2i-m+1\ p})\ \hat{f}_{k+2|j|+2i+2p+2}]\cos(kt + \frac{r+1}{2}\pi)$$

Durch eine Transformation des Index i entsteht daraus

$$J_{r+1\ m+1}\ f(t) - J_{r+1\ m}\ f(t)$$

woraus sich die Behauptung ergibt.

Aus Lemma 2 und Definition 3d) ergibt sich sofort

LEMMA 3. Für alle $r \in \mathbb{N}$ und $f \in P$ gilt

$$J_{r+1}\ f - \frac{1}{r}\cos J_r\ f + \frac{1}{r} J_r(\mathrm{id}.f) \in T_{r+1}$$

Die Lemmata 1 und 3 bilden genau die Voraussetzungen des Satzes 4, somit unterscheiden sich für $r \in \mathbb{N}$ und $f \in P$ die Funktionen $f \circ \cos$ und $J_r f^{(r)}$ nur durch ein (gerades) trigonometrisches Polynom vom Grad r.

3. DIE ZERLEGUNG DER FUNKTION $f \circ \cos$

Aus der Darstellung des Operators J_r ergibt sich auf natürliche Weise folgende Zerlegung von $f \circ \cos$:

DEFINITION 4. Für $r \in \mathbb{N}$ sei $\Omega_r := \bigcup_{m=0}^{r-1} \{m\} \times \mathbb{N}_0 \times \mathbb{N}^m$ und für $f \in P$ und $\omega = (m,p,j) \in \Omega_r$ sei weiters

$$f_\omega(t) := \frac{(-1)^m}{(r-1)!} \binom{r-1}{m} \sum_{k=1}^{\infty} (\Delta^{r-1+m} b_{k+2|j|-m\ p})\ \widehat{f^{(r)}}_{k+2|j|+2p} \cos(kt + \frac{r}{2}\pi)$$

Wegen $J_r f^{(r)} = \sin^r \sum_\omega f_\omega$ ist Problem 2A somit gelöst.

Für den Rest dieses Kapitels seien $n \geq r \in \mathbb{N}$ fix gewählt.
Im folgenden Lemma werden die Approximationsfehler der Funktionen f_ω abgeschätzt.

LEMMA 4. Sei a_k eine reelle Nullfolge mit $\Delta^h a_k \geq 0$, $h,k \in \mathbb{N}$ und sei weiters

$g(t) := \sum_{k=1}^{\infty} a_k \hat{f}_{k+j} \cos(kt + \frac{r}{2}\pi)$ mit $f \in P$ und $j \in \mathbb{N}_0$.

Dann gibt es ein $q_n \in T_{n-1}$ mit

$$\|g - q_n\| \leq \frac{8}{\pi} \sum_{l=0}^{\infty} \frac{1}{4l+1} a_{(4l+1)n} \|f\|$$

Beweis:

$$g(t) = \sum_{k=1}^{\infty} a_k \hat{f}_{k+j} \cos(kt + \frac{r}{2}\pi) =$$

$$= \frac{1}{\pi} \sum_{k=1}^{\infty} \int_0^{2\pi} a_k \cos(kt + \frac{r}{2}\pi - (k+j)u)\ f(\cos u)du =$$

$$= \int_0^{2\pi} [\frac{1}{\pi} \sum_{k=1}^{\infty} a_k \cos k(t-u)][\cos(ju - \frac{r}{2}\pi)\ f(\cos u)]du +$$

$$+ \int_0^{2\pi} [\frac{1}{\pi} \sum_{k=1}^{\infty} a_k \sin k(t-u)][\sin(ju - \frac{r}{2}\pi)\ f(\cos u)]du$$

Nach [2, p.175-178] gibt es A_n, $B_n \in T_{n-1}$ mit

$$\int_0^{2\pi} \left|\frac{1}{\pi} \sum_{k=1}^{\infty} a_k \cos ku - A_n(u)\right| du \leq \frac{4}{\pi} \sum_{l=0}^{\infty} \frac{(-1)^l}{2l+1} a_{(2l+1)n}$$

$$\int_0^{2\pi} \left|\frac{1}{\pi} \sum_{k=1}^{\infty} a_k \sin ku - B_n(u)\right| du \leq \frac{4}{\pi} \sum_{l=0}^{\infty} \frac{1}{2l+1} a_{(2l+1)n}$$

Mit $q_n(t) := \int_0^{2\pi} A_n(t-u)\cos(ju - \frac{r}{2}\pi)\ f(\cos u)du +$
$+ \int_0^{2\pi} B_n(t-u)\sin(ju - \frac{r}{2}\pi)\ f(\cos u)du$

folgt daraus die Behauptung.

Das nächste Lemma beweist, daß Lemma 4 zur Bestimmung der Konstanten $d_{\omega n}$ herangezogen werden kann.

LEMMA 5. Für $\omega := (m,p,j) \in \Omega_r$ sei

$$d_{\omega n} := \frac{8}{\pi} \frac{1}{(r-1)!} \binom{r-1}{m} \left| \sum_{l=0}^{\infty} \frac{1}{4l+1} \Delta^{r-1+m} b_{(4l+1)n+2|j|-m\ p} \right|$$

Dann gibt es zu jedem $f \in P$ ein $q_{\omega n} \in T_{n-1}$ mit $\|f_\omega - q_{\omega n}\| \leq d_{\omega n} \|f^{(r)}\|$

Beweis:

Für $p=0$ gilt $\Delta^h(\Delta^{r-1+m} b_{k+2|j|-m\,p}) = \Delta^{r-1+m+h} \frac{1}{k+2|j|-m+1}$,

für $p \geqq 1$ gilt $\Delta^h(-\Delta^{r-1+m} b_{k+2|j|-m\,p}) =$

$= \Delta^{r+m+h} \frac{1}{k+2|j|+2p-m-1} + \Delta^{r+m+1} \frac{1}{k+2|j|+2p-m}$

Weiters gilt für beliebige $h,k,j \in \mathbb{N}$

$$\Delta^h \frac{1}{k+j} = \frac{h!}{(k+j)(k+j+1)\dots(k+j+h)} > 0$$

Lemma 4 mit $g := \pm f_\omega$ ergibt die Behauptung.

LEMMA 6. $\sum_{\omega\in\Omega_r} d_{\omega n} < \frac{11}{3} (\frac{3}{2})^r \frac{(n-1)!}{(n+r-1)!}$

Beweis: Für jedes $k \in \mathbb{N}$ sei

$$c_k := \sum_{(m,p,j)\in\Omega_r} \frac{1}{(r-1)!} \binom{r-1}{m} |\Delta^{r-1+m} b_{k+2|j|-m\,p}|$$

Dann gilt $\sum_{\omega\in\Omega_r} d_{\omega n} = \frac{8}{\pi} \sum_{l=0}^{\infty} \frac{1}{4l+1} c_{(4l+1)n}$ und

$$c_k = \sum_{m=0}^{r-1} \frac{1}{(r-1)!} \binom{r-1}{m} \sum_{j\in\mathbb{N}^m} \Delta^{r-1+m}(b_{k+2|j|-m\,0} - \sum_{p=1}^{\infty} b_{k+2|j|-m\,p}) =$$

$$= \sum_{m=0}^{r-1} \frac{1}{(r-1)!} \binom{r-1}{m} \sum_{j\in\mathbb{N}^m} \Delta^{r-1+m} \frac{2}{k+2|j|-m+1} <$$

$$< \frac{2}{(r-1)!} \sum_{m=0}^{r-1} \binom{r-1}{m} \sum_{j_1=1}^{\infty} \dots \sum_{j_m=1}^{\infty} \Delta^{r-1+m} \frac{1}{k+2j_1-1+\dots+2j_m-1} \leqq$$

$$\leqq \frac{2}{(r-1)!} \sum_{m=0}^{r-1} \binom{r-1}{m} 2^{-m} \sum_{p_1=0}^{\infty} \dots \sum_{p_m=0}^{\infty} \Delta^{r-1+m} \frac{1}{k+p_1+\dots+p_m} =$$

$$= \frac{2}{(r-1)!} \sum_{m=0}^{r-1} \binom{r-1}{m} 2^{-m} \Delta^{r-1} \frac{1}{k} = 2(\frac{3}{2})^{r-1} \frac{1}{k(k+1)\dots(k+r-1)}$$

Daraus folgt für alle $l \in \mathbb{N}_0$

$$c_{(4l+1)n} < \frac{4}{3}(\frac{3}{2})^r \frac{1}{(4l+1)n} \cdot \frac{1}{((4l+1)n+1)\dots((4l+1)n+r-1)} \leqq$$

$$\leqq \frac{1}{4l+1} \frac{4}{3}(\frac{3}{2})^r \frac{1}{n(n+1)\dots(n+r-1)}$$

Damit erhält man

$$\sum_{\omega\in\Omega_r} d_{\omega n} = \frac{8}{\pi} \sum_{l=0}^{\infty} \frac{1}{4l+1} c_{(4l+1)n} < \frac{32}{3\pi} \sum_{l=0}^{\infty} (4l+1)^{-2} \left(\frac{3}{2}\right)^r \frac{(n-1)!}{(n+r-1)!} <$$

$$< \frac{11}{3} \left(\frac{3}{2}\right)^r \frac{(n-1)!}{(n+r-1)!}$$

Somit ist auch Problem 2B vollständig gelöst.

4. DIE KONSTANTE M_r

Die einzelnen Lemmata können nun zu folgendem Ergebnis zusammengefaßt werden:

SATZ 5. Für alle $r \in \mathbb{N}$ und $f \in C^r[-1,1]$ existiert für jedes $n \geq r$ ein Polynom p vom Grad höchstens n, sodaß für alle $x \in [-1,1]$ gilt:

$$|f(x) - p(x)| \leq \frac{11}{3} \left(\frac{3}{2}\right)^r \frac{(n-r)!}{n!} \sqrt{1-x^2}^{\,r} \, \|f^{(r)}\|$$

Die hier abgeleitete Konstante M_r ist natürlich nicht bestmöglich. Im Fall $r = 1$ besteht folgende

VERMUTUNG. Für alle $f \in C^1[-1,1]$ und alle $n \in \mathbb{N}$ existiert ein Polynom p vom Grad höchstens n, sodaß für alle $x \in [-1,1]$ gilt:

$$|f(x) - p(x)| \leq \tan \frac{\pi}{2(n+1)} \sqrt{1-x^2} \, \|f'\|$$

Wenn diese Vermutung richtig ist, so ist sie auch für jedes n bestmöglich. Dies zeigt folgendes

BEISPIEL. Für $n \in \mathbb{N}$ sei $f_n \in C[-1,1]$ absolut stetig mit $f_n(1) = 0$ und $f_n'(x) = \operatorname{sgn} \sin((n+1).\arccos x)$ f.ü.

Dann erhält man $\|f_n'\| = 1$, $|f(x)| \leq \tan \frac{\pi}{2(n+1)} \sqrt{1-x^2}$ und für die Punkte $x_k := \cos \frac{k\pi}{n+1}$, $k=0,\dots,n+1$, gilt $f_n(x_k) = (-1)^k \tan \frac{\pi}{2(n+1)} \sqrt{1-x_k^2}$

Für jedes Polynom p vom Grad höchstens n gibt es daher ein x_k, sodaß

$$|f(x_k) - p(x_k)| \geq \tan \frac{\pi}{2(n+1)} \sqrt{1-x_k^2}$$

LITERATURVERZEICHNIS

1. Fisher, S.O.: Best Approximation by Polynomials. Journal Approx.Th. 21 (1977), 43-59.

2. Schönhage, A.: Approximationstheorie. De Gruyter, Berlin 1963.

3. Sinwel, H.F.: Uniform Approximation of Differentiable Functions by Algebraic Polynomials. To appear in Journal Approx.Th.

4. Sinwel, H.F.: Konstanten in den Sätzen von Jackson und Timan. Reihe der Dissertationen der Universität Linz; VWGÖ, Wien 1981.

5. Sinwel, H.F.: Konstanten im Approximationssatz von Teljakowskii. Institutsbericht Nr. 186, Universität Linz, 1980.

6. Teljakowskii, S.A.: Two Theorems in the Approximation of Functions by Algebraic Polynomials. Mat.Sb. 70 (112) (1966), 252-265.

7. Timan, A.F.: Theory of Approximation of Functions of a Real Variable. Hindustan Publ. Comp., Delhi 1966.

MODELS FOR SMOOTH CURVE FITTING

Hans-Joachim Töpfer

Automatic control of plotters, tool machines, or robots is getting more and more important in industry. Most of the tools to be controlled require a smooth movement at a speed as high as possible, with respect to the materials of tool and workpiece. Often the movement is steered via several joints giving rise to a fairly high dimension of the control space. In this paper, parametric splines are used to handle such problems and a method for the optimal choice of the knot parameters is presented.

1. INTRODUCTION

Given n+1 points $p_0,\dots,p_n \in \mathbb{R}^m$ ($\mathbb{R}^m$ being the Euclidean m-dimensional space with inner product $(.,.)$ and the usual norm $\|.\|=\sqrt{(.,.)}$), find parameters $t_0<\dots<t_n \in \mathbb{R}$ and a smooth curve $x:[t_0,t_n]\to\mathbb{R}^m$, such that either

$$\text{(I)}\qquad x(t_i)=p_i \qquad (i=0,\dots,n)$$

or e.g.

$$\text{(F)}\qquad \sum_{i=0}^{n} \|x(t_i)-p_i\|^2=\min.$$

Properties (I) or (F) characterize the interpolative or least squares fit, respectively. For closed curves x(t) is considered to be (t_n-t_0)-periodic.

In the following we assume $x \in \left(C^p[t_o,t_n]\right)^m$, $(p \geq 1)$ and measure its smoothness by means of a given functional F, which has to attain an extreme value for the optimal x.

Functionals leading to an acceptable measure of smoothness often rely on technical or physical models. Examples of such models are:

Model 1: The travelling rocket model, with

$$F(x) := \int_{t_o}^{t_n} \|\ddot{x}(t)\| dt \,, \tag{1}$$

with the aim of minimizing the fuel consumption of a flying rocket;

Model 2: The nonlinear elasticity or geometric model, with

$$F(x) := \int_{t_o}^{t_n} (\ddot{x},\ddot{x})(t) dt \,, \quad \text{while } \|\dot{x}(t)\| \equiv 1 \,, \tag{2}$$

which minimizes the total bending energy of a beam represented by the curve x(t). In this case the parameter t is bound to be the arc length of the curve;

Model 3: The linear cubic spline model, with

$$F(x) := \int_{t_o}^{t_n} (\ddot{x},\ddot{x})(t) dt \,, \tag{3}$$

where the condition $\|\dot{x}(t)\| \equiv 1$ is dropped, which makes t a free parameter.

For fixed $t_o,\dots,t_n$ model (3) has been used by several authors. Ahlberg, Nilson and Walsh [1], for instance, propose to choose

$$t_o = 0 \quad \text{and} \quad t_{i+1} - t_i := \|p_{i+1} - p_i\|, \; (i=0,\dots,n-1).$$

Model (2) can be reformulated with independant t to give

$$F(x) := \int_{t_o}^{t_n} \frac{(\dot{x},\dot{x})(\ddot{x},\ddot{x}) - (\dot{x},\ddot{x})^2}{(\dot{x},\dot{x})^3}(t) dt \tag{2a}$$

unveiling the nonlinearity of the problem.

Different methods have been adopted to solve the associated variational problem in the two-dimensional case. Glass [5] and Woodford [10] use Euclidean coordinates and obtain the Euler equation in a necessarily complicated form, which is inappropriate for curves that are not monotone in at least one coordinate. Mehlum [7] and Lee and Forsythe [6] use the natural curve equation. Malcolm [8] gives a survey on existing results and methods and proposes a discrete natural spline approximation in connection with finite difference schemes. Bär [2] considers m-dimensional parametric splines and tries to approximately fulfill the condition $\|\dot{x}(t)\| \equiv 1$ by adding suitable conditions at the knots. Böhm [3] presents very interesting engineer's methods based on Bézier curves.

2. VARIATIONAL ANALYSIS

Let

$$T:=\left\{t=(t_0,\ldots,t_n)\in \mathbb{R}^{n+1} \;\middle|\; t_0<\ldots<t_n\right\}.$$

In the following, we consider the function spaces

$$X:=\left\{x:[t_0,t_n]\to\mathbb{R}^m \;\middle|\; x\in\left(C^1[t_0,t_n]\right)^m,\ \ddot{x}\in\left(L_2(t_0,t_n)\right)^m\right\}$$

and for given $t\in T$ and $p_0,\ldots,p_n\in\mathbb{R}^m$

$$X_0(t):=\left\{x\in X \;\middle|\; x(t_i)=p_i,\ (i=0,\ldots,n)\right\}.$$

For a given functional F the interpolation problem (I) leads to the following two steps:

Step I/1: For given $t\in T$ find $x^*\in X_0(t)$, such that

$$\bigwedge_{x\in X_0(t)} f(t):=F(x^*)\le F(x)\ .$$

Step I/2: Find $t^*\in T$, such that

$$\bigwedge_{t\in T} f(t^*)\le f(t)\ .$$

For the least squares fit problem (F) we have to modify the functional F, using a penalty parameter λ, to

$$\tilde{F}(x):=F(x)+\lambda\sum_{i=0}^{n}\|x(t_i)-p_i\|^2 .$$

We then have the following two steps:

<u>Step F/1</u>: For given $t\in T$ find $x^*\in X$, such that

$$\bigwedge_{x\in X} f(t):=\tilde{F}(x^*)\leq\tilde{F}(x) .$$

<u>Step F/2</u>: Find $t^*\in T$, such that

$$\bigwedge_{t\in T} f(t^*)\leq f(t) .$$

It is attractive to use standard variational techniques to solve the described partial problems, at least locally. However, this method turns out to be unsuitable for model (1), because of the singularity produced by the square root involved.

In the following, we consider Problem (I/3) in some detail and give the results for problems (F/3), (I/2), and (F/2).

2.1 <u>The interpolative cubic spline model (I/3)</u>

Consider an arbitrary $t\in T$ and the functional

$$F(x):=\int_{t_o}^{t_n}(\ddot{x},\ddot{x})(t)\,dt=\sum_{i=0}^{n-1}\int_{t_i}^{t_{i+1}}(\ddot{x},\ddot{x})(t)\,dt$$

We have to find an $x^*\in X_o(t)$, such that

$$\bigwedge_{x\in U(x^*)} F(x^*)\leq F(x) ,$$

where $U(x^*)$ is a certain neighborhood of x^*, and we have to establish conditions for the optimal choice of $t\in T$. The variation of $F(x)$ yields

$$\begin{aligned}\delta F(x)= &\int_{t_o}^{t_n}\left(\frac{d^4x}{dt^4},\delta x\right)(t)\,dt\\ &+[(\ddot{x},\ddot{x})-2(\dot{x},\dddot{x})](t_o^+)\,\delta t_o-[(\ddot{x},\ddot{x})-2(\dot{x},\dddot{x})](t_n^-)\,\delta t_n\\ &+\sum_{i=1}^{n-1}\Big[(\ddot{x},\ddot{x})-2(\dot{x},\dddot{x})\Big]_{t_i^-}^{t_i^+}\delta t_i\\ &+2\left(\dddot{x}(t_o^+),\delta x_o\right)-2\left(\dddot{x}(t_n^-),\delta x_n\right)\\ &+2\sum_{i=1}^{n+1}\left(\Big[\dddot{x}\Big]_{t_i^-}^{t_i^+},\delta x_i\right)\\ &-2\left(\ddot{x}(t_o^+),\delta\dot{x}_o\right)+2\left(\ddot{x}(t_n^+),\delta\dot{x}_n\right)\\ &-2\sum_{i=1}^{n-1}\left(\Big[\ddot{x}\Big]_{t_i^-}^{t_i^+},\delta\dot{x}_i\right).\end{aligned}$$

t_o and t_n and $x_o,\ldots,x_n$ considered to be fixed, we obtain the Euler equation

(4) $$\frac{d^4x}{dt^4}=0 \qquad \text{in } [t_i,t_{i+1}),\ (i=0,\ldots,n-1),$$

which can be integrated to give

(4a) $$x(t)\Big|_{[t_i,t_{i+1})}:=p_i+v_i(t-t_i)+\tfrac{1}{2}a_i(t-t_i)^2+\tfrac{1}{6}b_i(t-t_i)^3,$$

and the continuity condition for the second derivative

(5) $$\Big[\ddot{x}\Big]_{t_i^-}^{t_i^+}=0 \qquad (i=1,\ldots,n-1)$$

together with the natural boundary conditions

(5a) $$\ddot{x}(t_o^+)=\ddot{x}(t_n^-)=0.$$

The general properties of the space $X_o(t)$ provide for the continuity condition for the first derivative

(6) $$\left[\dot{x}\right]_{t_i^-}^{t_i^+}=0 \quad (i=1,\ldots,n-1)$$

and the interpolation condition

(7) $$x(t_i)=p_i \quad (i=0,\ldots,n),$$

which has already been used in (4a) to determine one constant of integration. These are the well-known properties of interpolating cubic splines.

Additionally we have the n-1 scalar equations

(8) $$\left[(\dot{x},\ddot{x})\right]_{t_i^-}^{t_i^+}=0 \quad (i=1,\ldots,n-1) \ ,$$

which give us the conditions required to determine the optimal $t_1,\ldots,t_{n-1}$. As $\dot{x}$ is continuous, (8) reads in fact

(8a) $$\left(\dot{x}(t_i),[\ddot{x}]_{t_i^-}^{t_i^+}\right)=0 \quad (i=1,\ldots,n-1),$$

which can be expressed, using the integration constants of (4):

(9) $$(v_i,b_i-b_{i-1})=0 \quad (i=1,\ldots,n-1).$$

This condition smooths the cubic spline further, as, in kinematic terms, x(t) considered as the path of a moving point, the projection of $\ddot{x}(t)$ in the direction tangential to the path has to be continuous in (t_0,t_n).

For given $t \in T$, the cubic parametric spline defined by (4) to (7) is not only a local, but also a global minimum point in $X_0(t)$ of the functional F (see e.g. [1]).

Condition (8) leads to a system of n-1 nonlinear equations for $t_1,\ldots,t_{n-1}$, which for all computed examples could be easily solved. However, further investigations, concerning conditions for existence and unicity of the solution, are necessary. In addition, one has to show that condition (8) characterizes the optimal t^*, mentioned as the aim of step I/2 of the minimizing procedure.

If other boundary conditions are to be applied, one can proceed in a very similar way. Such conditions may be:

1. At one or both ends of the curve, the direction of the tangent or the curvature is prescribed;
2. The curve is closed, i.e. $x(t)$ has to be (t_n-t_o)-periodic in each component, maintaining the continuity of $x^{(i)}(t)$ $(i=0,\dots,2)$ at $t_n \equiv t_o$.

We refrain from giving more details, here.

2.2 The least squares fit cubic spline model (F/3)

We have to consider the modified functional

$$\tilde{F}(x) := \int_{t_o}^{t_n} (\ddot{x},\ddot{x})(t)\,dt + \lambda \sum_{i=0}^{n} \Big(x(t_i)-p_i, x(t_i)-p_i \Big),$$

where λ is a penalty parameter which can be chosen in order to balance the two components of $\tilde{F}$.

A quite analogous analysis as in 2.1, where t_o and t_n but not the $x(t_i)$ are considered to be fixed, yields the following results:

The Euler equation (4) and the conditions (5), (6), and (8) remain unchanged. Condition (7) now becomes:

$$\text{(10)} \qquad \begin{aligned} x(t_o) &= p_o - \tfrac{1}{\lambda} x(t_o^+) \\ x(t_i) &= p_i - \tfrac{1}{\lambda} [x]_{t_i^-}^{t_i^+} \quad , \qquad (i=1,\dots,n-1) \\ x(t_n) &= p_n + \tfrac{1}{\lambda} x(t_n^-). \end{aligned}$$

Thus, the Euler equation can be integrated to give

$$\text{(11)} \qquad \begin{aligned} x(t)\Big|_{[t_o,t_1)} &= p_o - \tfrac{b_o}{\lambda} + v_o(t-t_o) + \tfrac{1}{2}a_o(t-t_o)^2 + \tfrac{1}{6}b_o(t-t_o)^3 \\ x(t)\Big|_{[t_i,t_{i+1})} &= p_i - \tfrac{1}{\lambda}(b_i - b_{i-1}) + v_i(t-t_i) \\ &\quad + \tfrac{1}{2}a_i(t-t_i)^2 + \tfrac{1}{6}b_i(t-t_i)^3. \end{aligned}$$

From (10) we see that for $\lambda\to\infty$ interpolation is attained.

Using (10), condition (8a) can be reformulated as

$$(12)\qquad \left(\dot{x}(t_i),p_i-x(t_i)\right)=0 \qquad (i=1,\ldots,n-1).$$

The meaning of (12) is quite simple: t_i has to be chosen, so that

$$(12a)\qquad p_i-x(t_i)\perp\dot{x}(t_i),$$

a very natural condition.

2.3 <u>The interpolative nonlinear elasticity model</u>

For the variational analysis of this model we have to apply the functional

$$\bar{F}(x)=\int_{t_o}^{t_n}\left\{(\ddot{x},\ddot{x})(t)+q(t)\,[(\dot{x},\dot{x})(t)-1]\right\}dt$$

with a Lagrange function q(t).

Standard techniques now yield the Euler equation, after one integration,

$$(13)\qquad \ddot{x}(t)-q(t)\dot{x}(t)=b_i,\quad t\in[t_i,t_{i+1}),\quad (i=0,\ldots,n-1).$$

Conditions (4),...,(8) remain unchanged.

In this case, t being the arc length of the curve x(t), t_n cannot be kept fixed, so there is one further condition:

$$(14)\qquad (\dot{x},\ddot{x})(t_n^-)=q(t_n^-),$$

whereas q(t) has to be determined in order to maintain $\|\dot{x}(t)\|\equiv 1$, thus making the differential equation (13) a nonlinear one.

Introducing $v:=\dot{x}$, turns (13) into

$$(15)\qquad \dot{v}(t)-q(t)v(t)=b_i \qquad t\in[t_i,t_{i+1}) \quad (i=0,\ldots,n-1)$$

and

$$(16) \qquad x(t) := p_o + \int_{t_o}^{t} v(s)ds \ .$$

The interpolation condition has to be replaced by

$$(17) \qquad \int_{t_i}^{t_{i+1}} v(s)ds = p_{i+1} - p_i \qquad (i=0,\ldots,n-1) \ .$$

Of course, the system of coupled nonlinear differential equations (13) or (15) together with conditions (4) to (8) or (17) is much more difficult to solve. Further experience will show, whether the advantages of this model versus the simpler linear cubic spline model, if there are any, are of practical relevance.

3. NUMERICAL RESULTS

A lot of examples have been computed. In the following, four of them are presented. Examples 2 and 3 are taken from [9] , example 4 from [10] . They shall show that the choice of the knot parameters may be critical. We found that, optimizing the knot parameters, leads to excellent results, even in cases, where parametric splines, usually, are said to be bad (see e.g. example 4, taken from [10]).

In all examples four different choices of the knot parameters t_i are given:

	Choice 1	Choice 2	Choice 3
$t_{i+1}-t_i$ =	$\|p_{i+1}-p_i\|_1$	$\|p_{i+1}-p_i\|_2^2$	$\|p_{i+1}-p_i\|_2$

Choice 4 means optimized knots.

The computations for choices 1 to 3 have been performed with the help of the FORTRAN-routine CUPIDF given in [9].

3.1 Example 1

Data			Knots			
No.	x_1	x_2	1	2	3	4
0	0	2.5	0	0	0	0
1	2.5	0	5.0	12.5	3.53553	2.83004
2	5.0	2.5	10.0	25.0	7.07107	7.77657
3	2.5	5.0	15.0	37.5	10.60660	10.60660
4	0	7.5	20.0	50.0	14.14214	13.43664
5	2.5	10.0	25.0	62.5	17.67767	18.38317
6	5.0	7.5	30.0	75.0	21.21320	21.21320
see figure			1.1	1.2	1.3	1.4

3.2 Example 2

Data			Knots			
No.	x_1	x_2	1	2	3	4
0	-0.5	2.5	0	0	0	0
1	-1.5	2.0	1.5	1.25	1.11803	1.14640
2	-1.5	1.0	2.5	2.25	2.11803	2.45321
3	-1.0	0.5	3.5	2.75	2.82514	3.32801
4	-0.5	0.5	4.0	3.0	3.32514	4.05100
5	0	1.0	5.0	3.5	4.03225	4.76354
6	1.0	2.0	7.0	5.5	5.44646	5.78960
7	2.5	3.0	9.5	8.75	7.24924	7.11695
8	3.5	3.0	10.5	9.75	8.24924	8.17488
9	3.5	1.5	12.0	12.0	9.74924	10.34471
10	1.5	1.0	14.5	16.25	11.81079	11.94691
11	0	0.5	16.5	18.75	13.39193	13.12758
12	-0.5	0	17.5	19.25	14.09904	13.81458
13	-0.5	-1.0	18.5	20.25	15.09904	14.95822
14	0	-1.5	19.5	20.75	15.80614	15.80370
15	1.0	-1.5	20.5	21.75	16.80614	16.80614
see figure			2.1	2.2	2.3	2.4

3.3 Example 3

Data			Knots			
No.	x_1	x_2	1	2	3	4
0	0	2.0	0	0	0	0
1	2.0	2.5	2.5	4.25	2.06155	1.94898
2	2.5	4.5	5.0	8.5	4.12311	4.16780
3	3.5	5.0	6.5	9.75	5.24114	5.38625
4	5.5	4.5	9.0	14.0	7.30269	7.58497
5	6.0	1.5	12.5	23.25	10.34407	10.57322
6	7.0	1.0	14.0	24.5	11.46211	11.66589
7	8.5	0.5	16.0	27.0	13.04325	13.20004
8	10.0	2.0	19.0	31.5	15.16457	15.16457
		see figure	3.1	3.2	3.3	3.4

3.4 Example 4

Data			Knots			
No.	x_1	x_2	1	2	3	4
0	0	0	0	0	0	0
1	2	-1.5	3.5	6.25	2.50000	3.69156
2	4	1.5	8.5	19.25	6.10555	8.16423
3	6	5.0	14.0	35.5	10.13668	12.03834
4	8	5.5	16.5	39.75	12.19823	15.11294
5	10	2.5	21.5	52.75	15.80378	18.64635
6	12	-4.5	30.5	105.75	23.08389	23.08389
		see figure	4.1	4.2	4.3	4.4

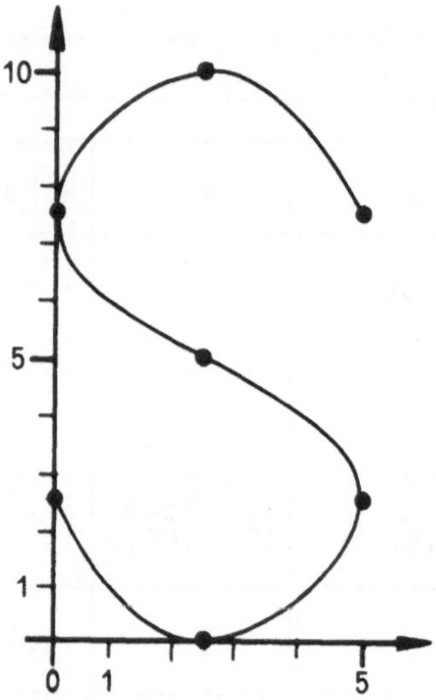

figure 1.1

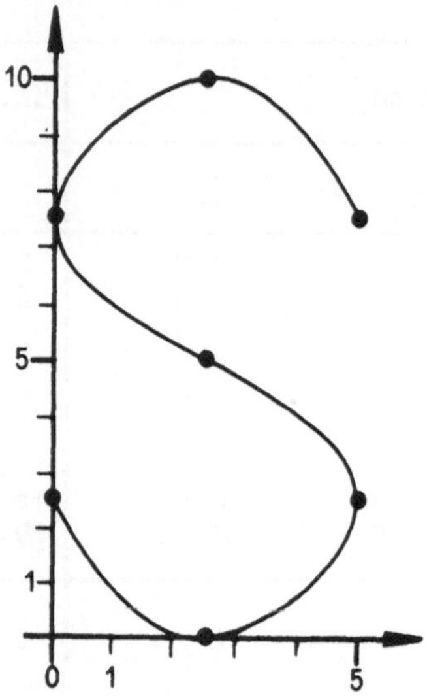

figure 1.2

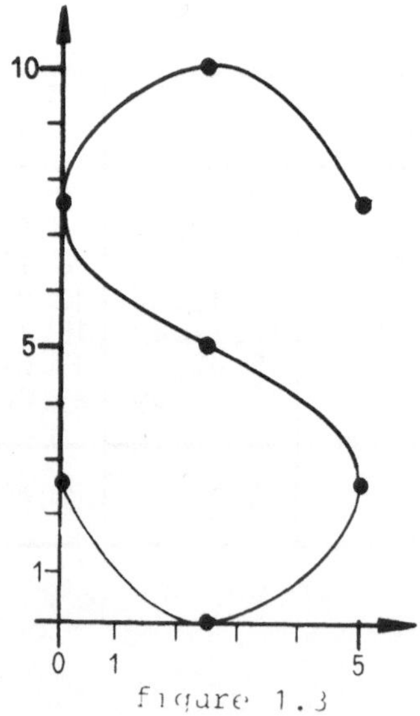

figure 1.3

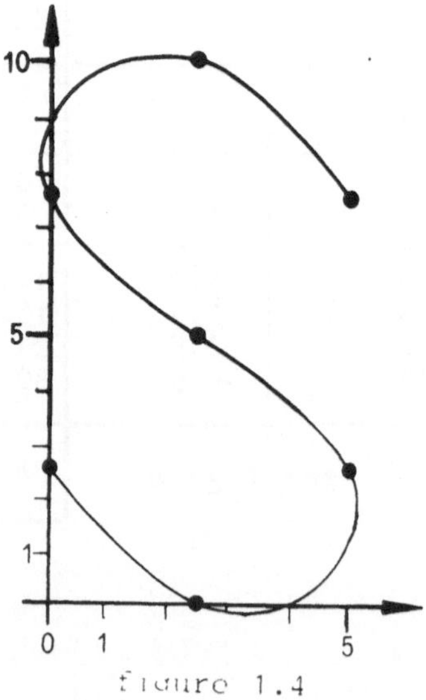

figure 1.4

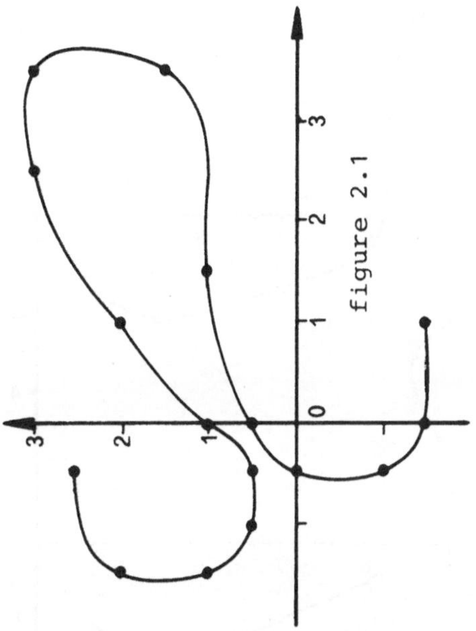

figure 2.1

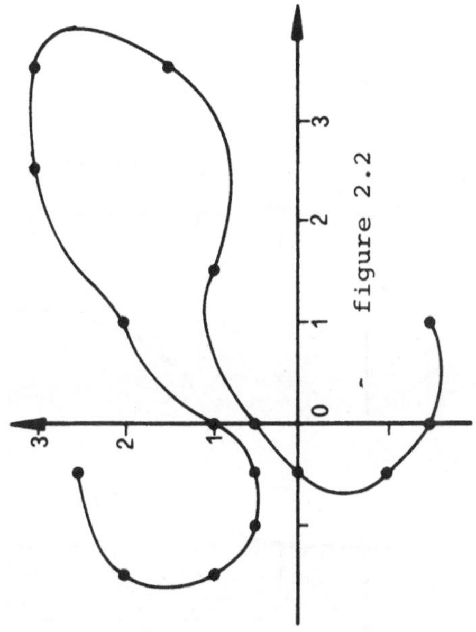

figure 2.2

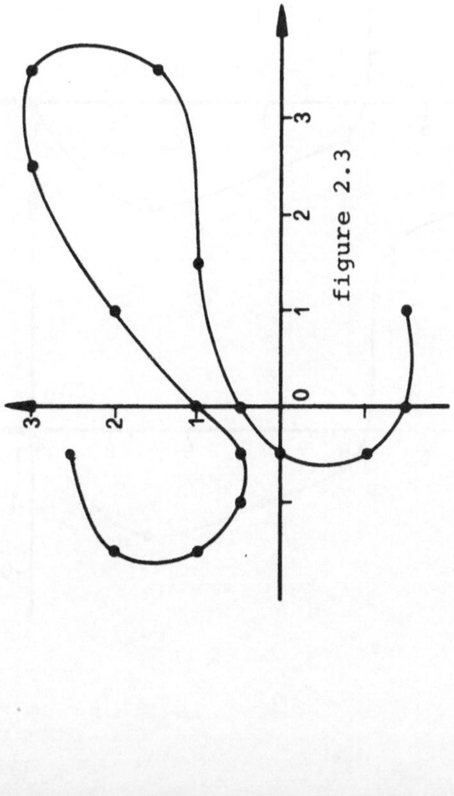

figure 2.3

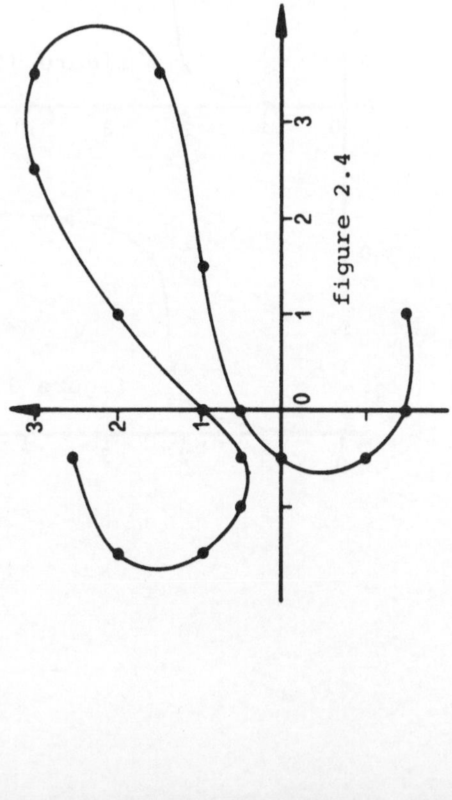

figure 2.4

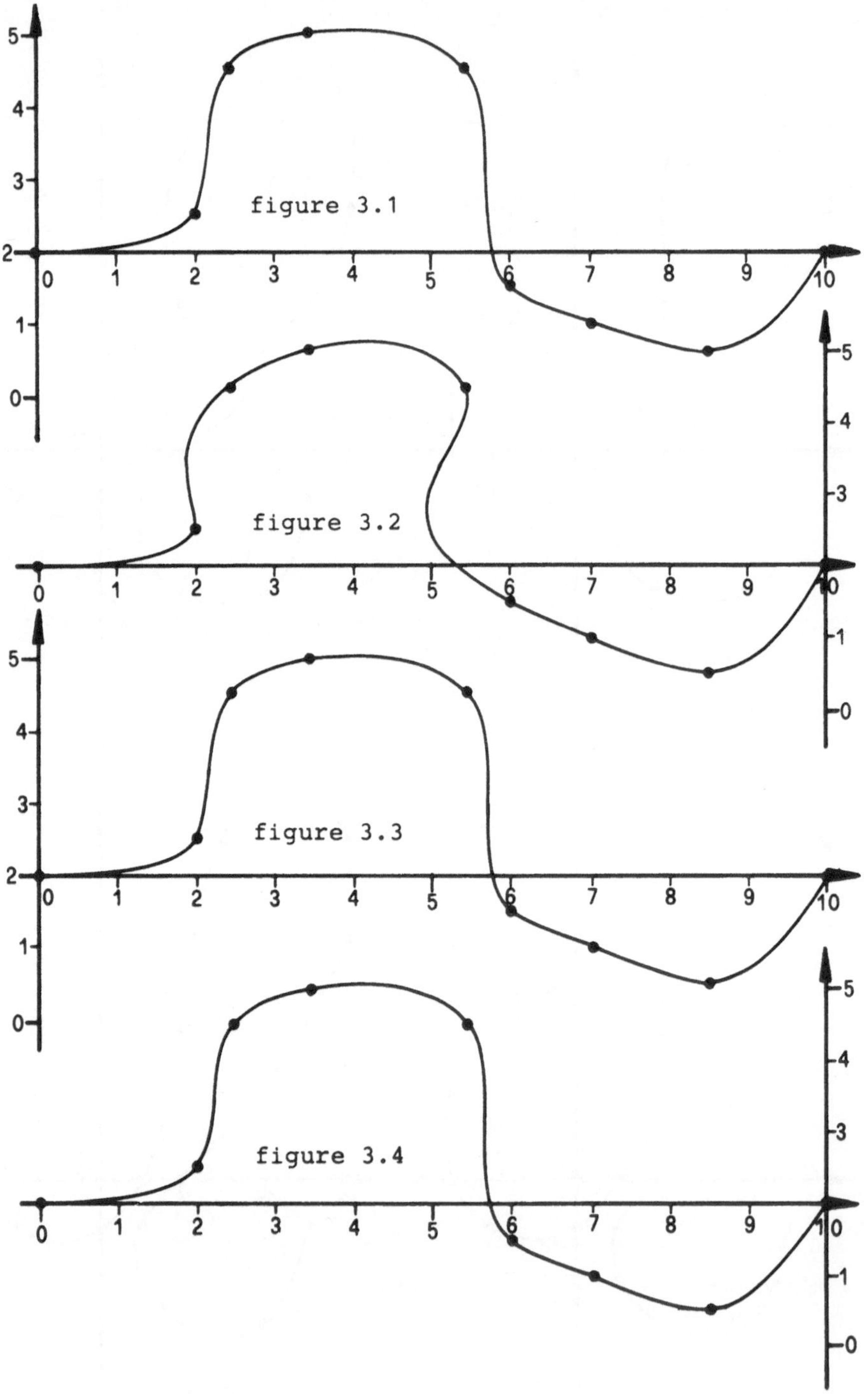

figure 3.1

figure 3.2

figure 3.3

figure 3.4

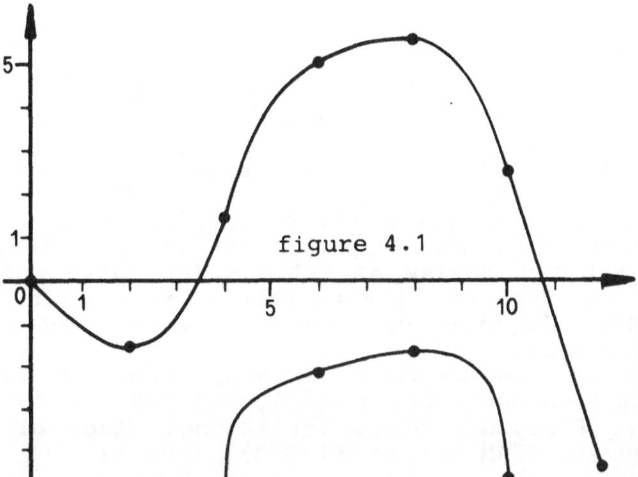

figure 4.1

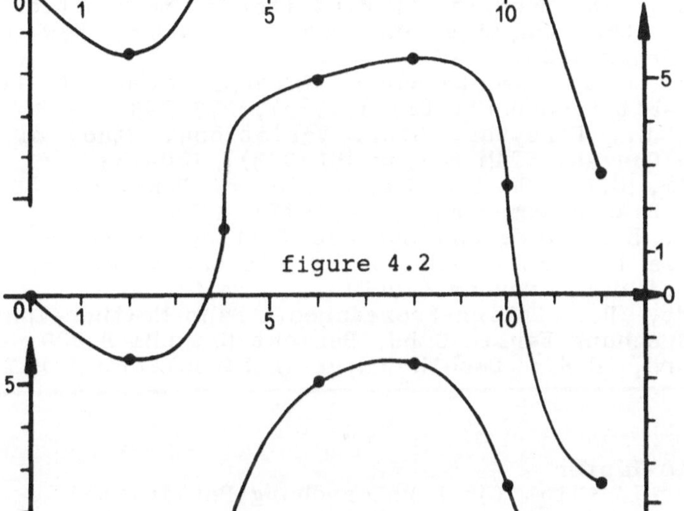

figure 4.2

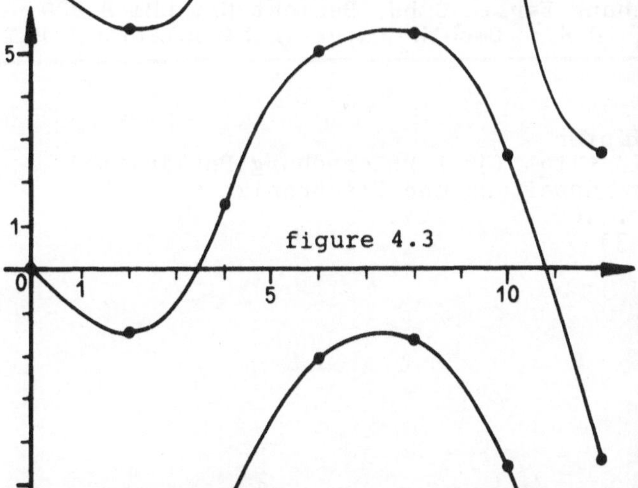

figure 4.3

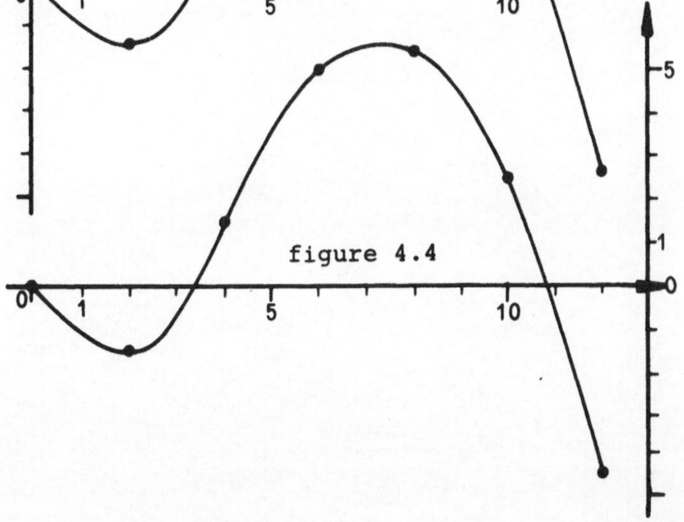

figure 4.4

REFERENCES

[1] Ahlberg, J.H., Nilson, E.N., Walsh, J.L.: The Theory of Splines and Their Applications. New York and London, Academic Press 1967

[2] Bär, G.: Parametrische Interpolation empirischer Raumkurven. ZAMM 57(1977), 305-314

[3] Böhm, W.: Cubic B-Spline Curves and Surfaces in Computer Aided Geometric Design. Computing 19(1977), 29-34

[4] Brodlie, K.W.: A Review of Methods for Curve and Function Drawing; in Brodlie, K.W.(editor): Mathematical Methods in Computer Graphics and Design. London, New York, Academic Press 1980, 1-37

[5] Glass, J.M.: Smooth-Curve Interpolation: A Generalised Spline-Fit Procedure. BIT 6(1966), 277-293

[6] Lee, E.H., Forsythe, G.E.: Variational Study of Nonlinear Spline Curves. SIAM Review 15(1973), 120-133

[7] Malcolm, M.A.: On the Computation of Nonlinear Spline Functions. SIAM J.Numer.Anal. 14(1977), 254-282

[8] Mehlum, E.: Curve and surface fitting based on variational criteria for smoothness. Ph.D. dissertation, Dept. of Math. Univ. of Oslo, Norway (1969)

[9] Schmidt, R.: Spline-Prozeduren. Hahn-Meitner-Institut für Kernforschung Berlin GmbH, Bericht Nr. HMI-B 220, 1976

[10] Woodford, C.H.: Smooth Curve Interpolation. BIT 9(1969), 69-77

Hans-Joachim Töpfer
Hahn-Meitner-Institut für Kernforschung Berlin GmbH
Bereich Datenverarbeitung und Elektronik
Glienicker Str. 100
D-1000 Berlin 39

SPLINE APPROXIMATION AS A TOOL FOR ESTIMATION

C.R. Traas

The functions, which describe the time evolution of a dynamic system, are approximated by means of splines. The unknown spline coefficients are estimated by processing noisy measurements of observable system variables, by collocating the system differential equation, and by using a-priori information. The arising overdetermined linear algebraic system in the spline coefficients is solved in a weighted least squares sense, with weights equal to the inverses of the appropriate standard deviations.

1. Introduction

A serious problem, occurring when state and parameter estimation methods are applied, is the problem of estimation inconsistency, the origin of which can be found in errors in the mathematical modelling of the dynamic system. In [2] a description is given of an investigation, the purpose of which was to determine the sensitivity of various estimation methods with respect to modelling errors. It was found that application of spline approximations in estimation, due to the flexibility of this type of approximation, considerably may improve the statistical consistency of the estimation process. In the present article the point of error sensitivity will not be treated (see [2] for this point), but emphasis will be placed upon the technique of using spline approximations in estimation. An illustration of the technique has been taken from the field of the space technique. It concerns the reconstitution of the attitude history of a spacecraft which experiences control torques and perturbing torques.

2. Differential equations

The system considered here is a linear second order system:

$$\left.\begin{aligned} \frac{d\omega}{dt} &= - A.(\omega-\omega_o) - B.\psi + M(t) \\ \frac{d\psi}{dt} &= \omega - \omega_o \end{aligned}\right\} \quad (1)$$

in which

ψ = the misalignment angle of the spacecraft w.r.t. the nominal attitude (i.e. the local zenith) (1-dimensional)

ω = the angular velocity of spacecraft rotation

ω_o = the nominal angular velocity of spacecraft rotation; $\omega - \omega_o$ = deviation of rotation velocity

$M(t)$ = perturbing troque (upper atmospheric effects)

A,B = constants of the dynamics (control torque).

The torques are expressed as torques per unit of moment of inertia of the spacecraft. The constants A and B are chosen such that, in the absence of a perturbing torque, the deviations of spacecraft attitude and angular velocity from their nominal values are nullified.

3. Measurements and a priori information

Measurements are provided at discrete times. They are of two types: angular velocity measurements by means of an on-board gyroscope, and attitude measurements by means of an on-board star sensor. Both measurement types are corrupted by measurement noise, which here is assumed to be white noise with known variance. In addition the gyro measurements contain an unknown contribution due to gyro drift. The expression for a gyro measurement z_{gyro} at some time t_ℓ is:

$$z_{gyro}(t_\ell) = \omega(t_\ell) + b_o + b_1 t_\ell + w_g(t_\ell) \quad (2)$$

in which $b_o + b_1 t_\ell$ is the gyrodrift, modelled here as a linear function of the time. The constants b_o and b_1 are unknown and should be estimated. $\omega(t_\ell)$ is the true angular velocity of the spacecraft at the time t_ℓ. The measurement white noise is represented by w_g. Properties of w_g are:

$$E\{w_g\} = 0$$

$$E\{w_g^2\} = \sigma_g^2 = \text{known variance.}$$

E = expectation operator.

The expression for a star sensor measurement z_{star} at some time t_r is:

$$z_{star}(t_r) = \phi_{star\ \nu} - \omega_o . t_r + \omega_o . w_s(t_r) \qquad (3)$$

in which

$\phi_{star\ \nu}$ = a celestial angular coordinate of the ν-th observed star

t_r = the star detection time

w_s = measurement white noise in star detection time

Properties of w_s are:

$$E\{w_s\} = 0$$

$$E\{w_s^2\} = \sigma_s^2 = \text{known variance.}$$

In the absence of white noise, $z_{star}(t_r)$ is just the attitude misalignment angle ψ of the spacecraft at the time t_r.

A priori information is knowledge about system parameters which is available in advance. Before any measurement is processed it is known, from some source, that $E\{\psi(t_o)\} = \psi$ (a priori), with a known standard deviation σ_ψ (a priori). The same information may be available for $\omega(t_o)$ and for the drift parameters b_o and b_1.

4. Spline approximations

The functions $\psi(t)$ and $M(t)$ are approximated in terms of normalized B-splines:

$$\left.\begin{aligned} \psi(t) &\simeq \sum_{i=-k+1}^{n} A_i B_{i,k}(t) \\ M(t) &\simeq \sum_{i=-k+3}^{n} M_i B_{i,k-2}(t) \end{aligned}\right\} \qquad (4)$$

in which the functions $B_{i,k}(t)$ and $B_{i,k-2}(t)$ are minimum support normalized basis spline functions of the orders k and k-2, respectively. The coefficients A_i and M_i are unknown and should be estimated, together with the drift parameters b_o and b_1, on the basis of the measurements, the a priori information and the collocation of the differential equation. For representing the gyro measurements and the differential equation, in terms of B-splines, derivatives of B-splines are required [1]. This leads to the expressions:

$$\left.\begin{aligned} \omega(t) &\approx \omega_o + (k-1) \sum_{i=-k+2}^{n} A_i^{[1]} B_{i,k-1}(t) \\ \dot{\omega}(t) &\approx (k-1)(k-2) \sum_{i=-k+3}^{n} A_i^{[2]} B_{i,k-2}(t) \end{aligned}\right\} \quad (5)$$

in which

$$A_i^{[1]} = \frac{A_i - A_{i-1}}{t_{i+k-1} - t_i}$$

$$A_i^{[2]} = \frac{A_i^{[1]} - A_{i-1}^{[1]}}{t_{i+k-2} - t_i}$$

where the t_i are the knots of the partition on which the B-splines are defined. The values of the B-splines $B_{i,k}$, up to and including order k, for a given t in the subinterval between the knots t_j and t_{j+1}, are computed recursively, using the following scheme [1]:

$$B_{i,q}(t) = \frac{t - t_i}{t_{i+q-1} - t_i} B_{i,q-1}(t) + \frac{t_{i+q} - t}{t_{i+q} - t_{i+1}} B_{i+1,q-1}(t)$$

where $q = 2,3,\ldots,k$

$i = j-q+1,\ldots,j$

$B_{j,1} = 1, \quad B_{j+1,1} = 0$

$B_{i,q} = 0$ when $i > j$ or $i < j-q+1$

5. System of linear equation

The measurement equations (2) and (3) give rise to the following equations in the spline coefficients and drift parameters:

$$\left.\begin{aligned}&(k-1)\sum_{i=-k+2}^{n} A_i^{[1]} B_{i,k-1}(t_\ell) + b_o + b_1 t_\ell = z_{gyro}(t_\ell) - \omega_o,\\&\sum_{i=-k+1}^{n} A_i B_{i,k}(t_r) = z_{star}(t_r)\end{aligned}\right\} \quad (6)$$

The a priori information leads to the following relations:

$$\left.\begin{aligned}&(k-1)\frac{A_{-k+2}-A_{-k+1}}{t_1-t_o} = \omega(\text{a priori}) - \omega_o\\&A_{-k+1} = \psi(\text{a priori})\\&b_o = b_o(\text{a priori})\\&b_1 = b_1(\text{a priori})\end{aligned}\right\} \quad (7)$$

The simple form of the first two equations in (7) is due to the k-fold multiplicity of the boundary knot t_o.
The equations (6), with the indices ℓ and r running over all gyromeasurements and starmeasurements, respectively, together with the equations (7), form an overdetermined linear system which can be solved, in weighted least squares sense, for the spline coefficients A_i (i=-k+1,....,n), and the gyrodrift parameters b_o and b_1. The weight assigned to each of these equations in this solution process, equals the inverse of the appropriate standard deviation.

Substitution of (4) and (5) into the first relation of (1) gives:

$$\begin{aligned}(k-1)(k-2)\sum_i A_i^{[2]} B_{i,k-2}(t) &=\\ &= -A.(k-1)\sum_i A_i^{[1]} B_{i,k-1}(t) - B\sum_i A_i B_{i,k}(t) + \qquad (8)\\ &+ \sum_i M_i B_{i,k-2}(t)\end{aligned}$$

The relation (8) will of course, in general, not indentically be true for all t (if it were, then the approximation $\psi(t) = \sum_i A_i B_{i,k}(t)$ would be an exact solution to (1)). We may require, however, that (8) is true for a finite number of points t_c, the collocation points. Choosing a number of collocation points equal to the number of spline M_i-coefficients, and distributed suitably over the time interval considered, a linear system arises form which the M_i

can be solved, using the known A_i coefficients. A weight for these equations is relevant only if computation of the covariance matrix, associated with the computed M_i coefficients, is required. A suitable weight, however is not easily found, since it depends on a statistical measure of the approximation errors of the spline representations. If error estimates are available, they generally are given in the form of upper-bound expressions, and in many cases these bounds may be very course. The best way to get a suitable measure for approximation errors is by experimenting with the positions and the number of knots, or by simulating the system realistically and comparing approximate results with the exact solution.

6. Results

For a specific choice of the relevant parameters numerical results have been produced. These results confirm the flexibility of the present approach, in the sense that no severe a priori restrictions have to be made with respect to the class of functions to which the perturbing torque M(t) belongs in the considered time interval. This is of great importance since the correct physical modelling of the perturbing torque is a very difficult matter due to lack of knowledge about properties and behaviour of the upper atmosphere. Some of the results of one specific simulation are presented in the figures. The time interval considered is from 0 to 6000 seconds. The perturbing torque as a function of time is given in figure 1, and the corresponding attitude history of the spacecraft in figure 2.

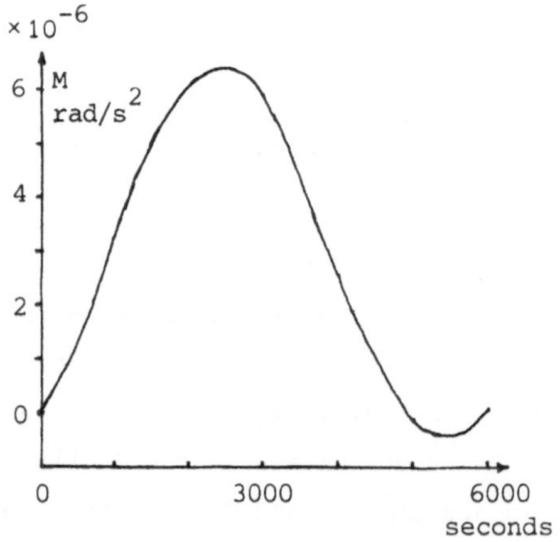

Fig. 1: Perturbing torque

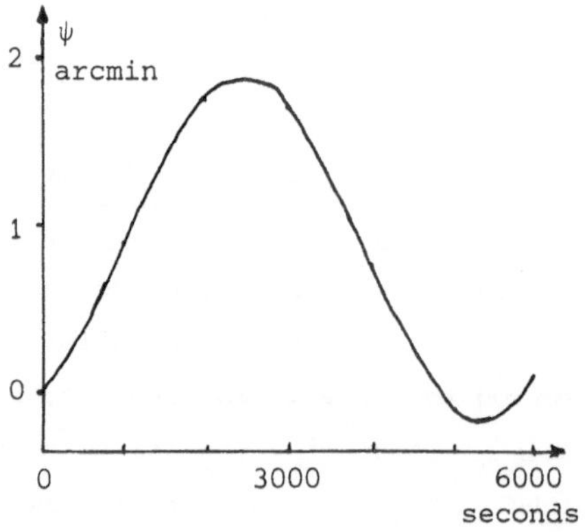

Fig. 2: Attitude history

The estimation errors, as far as M and ψ are concerned, are given in the figures 3 and 4, respectively. Each of these figures shows 3 results, namely for spline approximations of the orders 4, 5 and 6, respectively, with respect to $\psi(t)$ (and thus the orders 2, 3 and 4 with respect to $M(t)$).

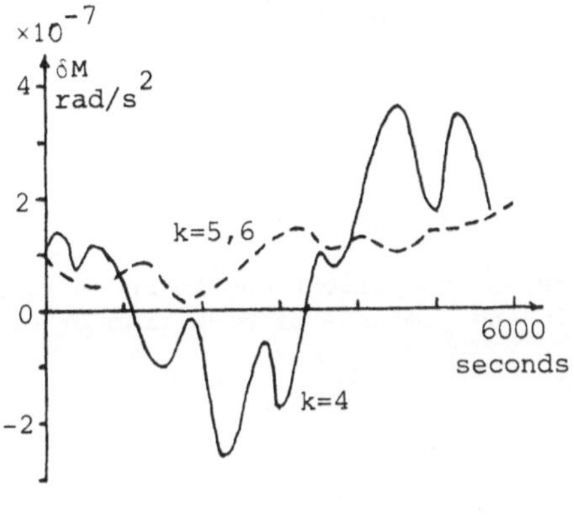

Fig. 3: Estimation errors in perturbing torque

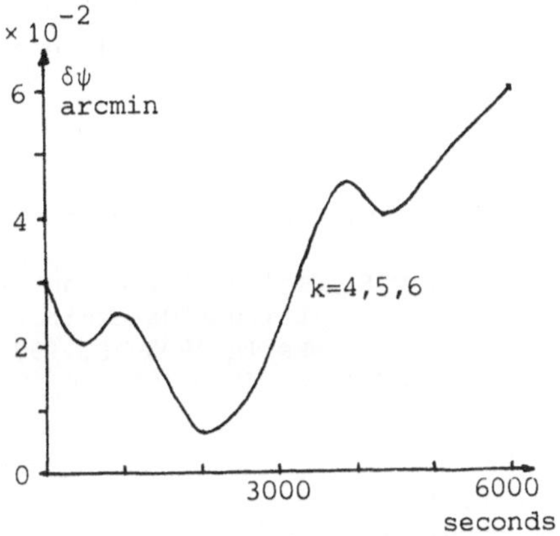

Fig. 4: Estimation errors in attitude angle

The results with 5th and 6th order splines (for $\psi(t)$) are mutually very close, but in the case of fourth order splines the errors in the estimated torque are larger. It may be expected that this is a consequence of the simple (namely piece-wise linear) approximation used for $M(t)$ in this case.

The numerical results have been prepared with the following parameter values:

- number of starsensor measurements: 15
- number of gyromeasurements: 300
- standard deviation of starsensor measurement noise: $\sigma_s = 0.04$ sec
- standard deviation of gyro measurement noise: $\sigma_g = 0.2 \times 10^{-7}$ rad/sec
- the gyro drift is $0.5 \times 10^{-7} + 1.5 \times 10^{-10} \times t$ rad/sec.

The number of knots used in the partition is 10. The boundary knots have multiplicity equal to the order k of the splines used for approximating ψ.

The number of collocation points is 10 in the case of $k = 4$. These points are chosen to coincide with the knots.

When $k = 5$, the number of collocation points is 11, and they are chosen to coincide with the boundary knots and with the middles of the subintervals.

When $k = 6$, the number of collocation points is 12, and they are chosen to coincide with the knots and with the middles of the first and the last subintervals.

The estimates of the gyro drift coefficients are $b_0 = 0.5005 \times 10^{-7}$ and $b_1 = 1.480 \times 10^{-10}$, in all spline order cases.

7. References

[1] de Boor, C.: A practical guide to spline analysis, 1978.

[2] Traas, C.R.: Digital filtering methods, with applications to spacecraft attitude determination in the presence of modelling errors NLR TR 79039 L, 1979.

Dr. C.R. Traas
Twente University of Technology
Department of Applied Mathematics
P.O. Box 217
7500 AE Enschede, The Netherlands

A GLOBALLY CONVERGENT METHOD FOR (CONSTRAINED) NONLINEAR CONTINUOUS L_1 APPROXIMATION PROBLEMS

G. Alistair Watson

Let $f,g : X \times R^n \to R$ be continuously differentiable functions of their parameters, and consider the problem

$$\text{find } a \in R^n \text{ to minimise } ||f(x,a)||$$
$$\text{subject to } g(x,a) \leq 0 , x \in X ,$$

where $X \subset R^N$ is a Cartesian product of closed intervals, and the norm is the L_1 norm defined on X. A globally convergent method is given for this problem, based on the use of an exact penalty function, and capable of converging at a second order rate if exact second derivative information is also available. For the particular case of the unconstrained problem, and when X is an interval of the real line, an algorithm is presented, and the effectiveness of this is illustrated by the numerical solution of a number of nonlinear problems.

1. INTRODUCTION

Let $X \subset R^N$ be a Cartesian product of closed intervals, and let $f,g : X \times R^n \to R$ be continuously differentiable functions of their parameters. Then the problem

$$\text{find } a \in R^n \text{ to minimise } ||f(x,a)|| = \int_X |f(x,a)|dx \tag{1.1}$$
$$\text{subject to } g(x,a) \leq 0 , x \in X$$

is an example of a constrained L_1 approximation problem. The problem of L_1 approximation of continuous functions does not occur as frequently as, for example, Chebyshev approximation. However, constrained L_1 problems arise naturally in applications; in particular, the special case of (1.1) when $g = f$ (the one-sided problem) arises, for example, in the approximation of the density of a probability distribution ([3], [10]) and in the approximate solution of weakly singular integral equations ([1], [9]). There is some simplification in (1.1) with such problems, for the modulus signs may be discarded, and the problem dealt with by techniques appropriate to more conventional semi-infinite programming problems. When f cannot be assumed to be of constant sign in X, then special treatment of the objective function is necessary.

We present here a globally convergent method for (1.1) based on the use of an associated exact penalty function. The approach is capable of giving convergence to a stationary point of (1.1), which we now define. We will use ϕ_j to denote $\frac{\partial f}{\partial a_j}$, ψ_j to denote $\frac{\partial g}{\partial a_j}$, $j = 1,2,\ldots,n$, and we will suppress the explicit dependence of quantities on their parameters when no confusion is likely.

DEFINITION 1 A point $a \in R^n$ is a stationary point of (1.1) if it is feasible in the constraints of (1.1), and in addition, there exist points $x_i \in X$ with $g(x_i,a) = 0$, $i = 1,2,\ldots,t$, non-negative numbers λ_i, $i = 1,2,\ldots,t$ and $v \in \partial||f||$ (the sub-differential of $||f||$) such that

$$< \phi_j, v > + \sum_{i=1}^{t} \lambda_i \psi_j(x_i) = 0 , \quad j = 1,2,\ldots,n . \tag{1.2}$$

Under an appropriate constraint qualification, the analysis given, for example, in [2] shows that these conditions are necessary for a to solve (1.1). Considerable simplification may be obtained if the zeros of f in X have measure zero, for then

$$< \phi_j, v > = \int_X \phi_j \text{ sign } (f) \, dx , \quad j = 1,2,\ldots,n , \tag{1.3}$$

independent of v. We will assume in what follows that (1.3) holds at all points of interest; the jth element of (1.3) is then just the partial derivative of $||f||$ with respect to a_j.

In the next section, we describe the exact penalty function approach which is proposed for the calculation of stationary points of (1.1). In Section 3, we restrict attention to the unconstrained problem, for the case when X is an interval of the real line, and so can concentrate on the special treatment of the objective function. An algorithm is given for this problem, which generalises that given in [11] for the linear case, and which does not require the calculation of second derivatives of f. The method is illustrated in Section 4 by the solution of a number of nonlinear L_1 approximation problems.

2. AN EXACT PENALTY FUNCTION APPROACH

We will assume now that g is a twice continuously differentiable function of x, and will denote by $\nabla_x g$ and $\nabla_x^2 g$ respectively the vector of partial derivatives, and matrix of second partial derivatives of g with respect to x. Let $\eta > 0$ be prescribed, and denote by $E(a)$ the set of local maxima of g in X with $g(x,a) \geq -\eta$. Define $B \subset R^n$ as the set of points a such that

(i) $E(a)$ contains a finite number of points $x_1, x_2, \ldots, x_p$,

(ii) there exists a fixed number $\varepsilon > 0$ such that

(a) $||\nabla_x g(x_i,a)|| \geq \varepsilon$ if $x_i \in E(a)$ is a boundary point of X

(b) $\det[\nabla_x^2 g(x_i,a)] \leq -\varepsilon$ if $x_i \in E(a) \cap \text{int}(X)$.

Let the boundary points of $E(a)$ be x_i, $i = \bar{p}+1, \ldots, p$. Then for any $a \in B$, we may define the function

$$P(a,\theta) = ||f(a)|| + \theta \sum_{i=1}^{p} [h_i(a)]_+ , \qquad (2.1)$$

where

$$\nabla_x g(x_i,a) = 0 , \quad i = 1,2,\ldots,\bar{p} , \qquad (2.2)$$

$$h_i(a) = g(x_i(a),a) , \quad i = 1,2,\ldots,\bar{p} ,$$

$$h_i(a) = g(x_i,a) , \quad i = \bar{p}+1,\ldots,p ,$$

and θ is a positive constant. Our assumptions mean that the implicit function theorem may be applied to (2.2) to give the points x_i, $i = 1,2,\ldots,\bar{p}$ as differentiable functions of a. It follows that $P(a,\theta)$ is a locally Lipschitz function, and so necessary conditions for a local minimum correspond to the existence of a zero generalized gradient [2].

THEOREM 2.1 Let $a \in B$ minimize $P(a,\theta)$. Then there exist numbers w_i, $0 \leq w_i \leq 1$, $i = 1,2,\ldots,p$ with $w_i = 1$ if $h_i > 0$, $w_i = 0$ if $h_i < 0$, and $v \in \partial||f||$ such that

$$< \phi_j, v > + \theta \sum_{i=1}^{p} w_i \psi_j(x_i) = 0 , \quad j = 1,2,\ldots,n . \qquad (2.3)$$

Comparison of (2.3) with (1.2) shows that P fulfils the role of an exact penalty function for (1.1) : any local minimiser which is feasible in the constraints of (1.1) is a stationary point. Let $a \in B$ be an approximation to the solution of (1.1) and let

$$c_j = \langle \phi_j, v \rangle , \quad j = 1,2,\ldots,n .$$

Then the minimisation of P may be achieved, for example, by a method based on the following iteration.

1. Find $d \in R^n$ to minimise $d^T c + \frac{1}{2} d^T H d$ (2.4)

   subject to $g(x_i) + d^T \psi(x_i) \le 0$, $i = 1,2,\ldots,p$,

   where H is a given $n \times n$ positive definite matrix.

2. Choose γ as the maximum of the sequence $\{1,\beta,\beta^2,\ldots\}$, $0 < \beta < 1$ so that

$$Q(a,\gamma,\theta) \equiv \frac{P(a,\theta) - P(a + \gamma d,\theta)}{-\gamma\, G(\theta)} \ge \sigma , \qquad (2.5)$$

   where $0 < \sigma < 0.5$, and $G(\theta)$ is the generalized directional derivative of $P(a,\theta)$ at a in the direction d, given by

$$G(\theta) = d^T c + \theta \left(\sum_{i \in I_0} [d^T \psi(x_i)]_+ + \sum_{i \in I_+} d^T \psi(x_i) \right) ,$$

   with $I_0 = \{i : x_i \in E , \ g(x_i,a) = 0\}$,

   $I_+ = \{i : x_i \in E , \ g(x_i,a) > 0\}$.

3. Replace a by $a + \gamma d$.

   Under the conditions

   (i) $\theta \ge$ {maximum of the Lagrange multipliers defined at solutions to (2.4)}

   (ii) the points generated remain in a bounded region of B in which f and g have bounded second derivatives,

the limit points of the above iteration may be shown to be stationary points of (1.1). The analysis is similar to that given in [12] for the usual semi-infinite case, because of the assumption (1.3).

The subproblem (2.4) is based on the SOLVER method (see, for example [5]), and a possible difficulty with this is that the constraints may be inconsistent, so that no d exists; also it is possible that the solutions d do not remain bounded. To avoid difficulties of this kind, an alternative subproblem is suggested (for the finite case) by Fletcher [4] ; for our purposes, this means replacing (2.4) by

1a Find $d \in R^n$ to minimise $d^Tc + \frac{1}{2}d^THd + \theta \sum_{i=1}^{p} [g(x_i) + d^T\psi(x_i)]_+$

$$\text{subject to} \quad ||d|| \leq \delta , \tag{2.6}$$

where the norm on R^n is the L_∞ norm, and δ is a prescribed positive number.

This problem always has a solution, and the possibility of unboundedness of d does not arise. Suitable choice of δ can ensure that the iterations remain in a 'trust' region in which the approximations to the objective function and constraints are satisfactory ones. A basic convergence result may again be obtained under condition (ii) above, provided that δ is chosen in a suitable way and also that the matrices H remain bounded (see, for example, [12]) ; however, feasibility in (1.1) of limit points is not automatically guaranteed, and the role of θ here is to be big enough to force the extra terms in the objective function of (2.6) to be zero in the limit.

We turn now to the choice of H at each iteration. By analogy with finite programming problems, we would like H to be equal to (or approximately equal to) the matrix of second derivatives at the current point with respect to the components of an appropriate Lagrangian function. Thus if g is a twice continuously differentiable function of a, we can take H to be the matrix

$$G + \sum_{i=1}^{p} \lambda_i \nabla^2 h_i(a) \tag{2.7}$$

where λ_i, $i = 1,2,\ldots,p$ are Lagrange multiplier estimates, and G takes account of second derivative information about the objective function. If the matrix (2.7) is not positive definite, then some positive definite approximation is required. On the other hand if G is the exact matrix of second partial derivatives of the objective function, and H is defined by (2.7), then it is possible to establish a precise connection between the above methods and Newton's method applied to (1.2) ([12] ; see also [7], [8]).

3. AN ALGORITHM FOR THE UNCONSTRAINED PROBLEM

We consider now the special case of (1.1) when no constraints are present, and when X is an interval of the real line. We will assume also that at all points of interest, f has a finite number of zeros in X. If these zeros are denoted by $z_1, z_2, \ldots, z_m$, and it is further assumed that they are simple, then the matrix of second partial derivatives of $||f||$ with respect to the components of $\mathbf{a}$ is given by

$$C + A^T D^{-1} A \tag{3.1}$$

where

$$C_{ij} = \sum_{k=1}^{m+1} (-1)^k \int_{z_{k-1}}^{z_k} \frac{\partial^2 f}{\partial a_i \partial a_j} dx \ , \ i,j = 1,2,\ldots,n \ ,$$

$$A_{ij} = \phi_j(z_i) \ , \ i = 1,2,\ldots,m \ , \ j = 1,2,\ldots,n \ ,$$

$$D_{ij} = \delta_{ij} \ |\tfrac{1}{2} f'(z_i,a)| \ , \ i,j = 1,2,\ldots,m \ ,$$

$X = [z_0, z_{m+1}]$ and the prime denotes differentiation with respect to x. We would like, therefore, to take H to be a positive definite approximation to the matrix (3.1), and propose the following method of choosing H, similar to that of [11] for the linear problem.

case (a) $\quad m = 0$ (no zeros) or D singular

$$H = I \tag{3.2}$$

case (b) $\quad D$ nonsingular, rank $(A) = n$

$$H = A^T D^{-1} A \tag{3.3}$$

case (c) $\quad D$ nonsingular, rank $(A) < n$

$$H = A^T D^{-1} A + \mu I \ , \quad \mu > 0 \ . \tag{3.4}$$

The advantages of using these rules are that (i) no second derivatives of f are required (ii) H is always positive definite (iii) the calculation for d, which corresponds to the solution of the linear system

$$Hd = -c \ ,$$

may be achieved in a stable manner, through e.g. factorization of $D^{-\frac{1}{2}}A$ into the product of an orthogonal matrix and an upper triangular matrix. On the other hand, since exact second derivatives are not used, the second order convergence rate which can occur if (3.1) is used will usually be lost. A more serious problem in solving nonlinear problems is the greater difficulty in obtaining good initial approximations, and this makes it necessary to be rather more careful in devising a satisfactory overall strategy (compared to solving linear problems) in order to lessen the risk of very slow convergence far from the solution. Many of the rules given below, and also the values of the various tolerances, multipliers and bounds used, are nevertheless somewhat arbitrary. Variations on these were tried which gave improved results in some cases, but no set of rules used gave consistently better results over the range of problems which was solved.

The lack of a good initial approximation means that for most problems we are attempting to start from an approximation with few (or no) interpolation points, and to move to one with at least n. This suggests that the incorporation of some strategy which takes account of this would be useful, and it was in fact found to be beneficial in practice to modify the step length test (2.5), so that in addition to (2.5), we also require that the number of interpolation points in the next approximation is not reduced. If the result of this is a step $\gamma < \gamma_1$, where γ_1 is a prescribed lower bound, then we proceed according to the following rules, which differ to take account of the nature of the current search direction (as determined by the choice of H) in each case.

case (a) Choose γ using (2.5) only ; if this gives $\gamma < \gamma_2$, then terminate.

case (b) Set μ to $M\mu_0$ and recalculate H from (3.4); calculate γ using (2.5) only; continue to use (3.4) for subsequent calculations of H.

case (c) If $M\mu > \mu_2$ (where μ is the current value), set $H = I$; otherwise replace μ by $M\mu$ and recalculate H from (3.4) ; in either case calculate γ using (2.5) only.

Finally, we give the rules used for the adjustment of μ. The strategy is motivated by an attempt to keep good agreement between the actual and predicted reduction in P (= $||f||$) as measured by Q, and also to keep the actual step length as close to unity as possible. Initially, μ is set to μ_0, and reset to μ_0 if a steepest descent step (case (a)) is taken at any time. After a step with H given by (3.4), we proceed as follows:

if $\gamma = 1$ and $Q > 0.5$, reset μ to $\mu/4$
if $2^{-3} \le \gamma < 1$, reset μ to 4μ
if $\gamma < 2^{-3}$, reset μ to 16μ .

In addition, μ is subject to the bounds

$$\mu_1 \le \mu \le \mu_2$$

and is set to an extreme value if this value is liable to be exceeded.

The parameters used in the numerical results quoted in the next section were

$$\gamma_1 = \gamma_2 = 2^{-6} , M = 64 ,$$
$$\lambda_0 = 1 , \lambda_1 = 4^{-4} , \lambda_2 = 4^5 ,$$

and all the results were obtained from an Algol program run on the DEC-10 computer at the University of Dundee, which gives about 8 decimal places for single length working. The algorithm was (successfully) terminated on satisfaction of the condition

$$c^T c < 10^{-7} . \tag{3.5}$$

4. NUMERICAL RESULTS

In all the examples quoted, a^o is the initial approximation, a^* the final approximation obtained when (3.5) is satisfied, and m is the number of zeros of f . The number N denotes the number of iterations (number of times (d,γ) is calculated), and N_e denotes the number of times $||f||$ is evaluated. The norm was obtained by using Simpson's rule in each subinterval between pairs of zeros, with interval subdivision until the values returned agreed to 5 decimal places. In the table, 'case' refers to the choice of H , and γ is the step length.

Example 1

$$f(x,a) = a_1^2 + 2 a_1 a_2 x^2 - \sin x , \quad X = [0,2]$$

$$a^o = (1,1)^T, \ ||f(a^o)|| = 1.206\ 968 .$$

$$a^* = (-0,482\ 148 , -0.778\ 177)^T , N_e = 8 .$$

The details are given in Table 1.

iteration	m	case	γ	$\|\|f\|\|$
1	0	(a)	0.5	0.508 712
2	1	(c)	1	0.448 879
3	1	(c)	1	0.075 837
4	2	(b)	1	0.065 494
5	2	(b)	1	0.065 149
6	2	(b)	1	0.065 148

Table 1

Example 2

$$f(x,a) = a_1 + a_2 e^{a_3 x} - 5 - 6e^{2x} - 2 \sin(4x) \ , \ X = [0,1]$$

$a^o = (1,1,1)^T \qquad ||f(a^o)|| = 22.275 \qquad m = 0$

$a^* = (1.782\ 979,\ 10.039\ 63,\ 1.493\ 393)^T, \ ||f(a^*)|| = 0.300\ 725$

$m = 3 \ , \ N = 15 \ , \ N_e = 28 \ .$

Example 3

$$f(x,a) = \frac{a_1 + a_2 x}{1 + a_3 x} - \phi(x) \ , \ X = [0,1]$$

(a^o was obtained by interpolation at $x = 0.1, 0.5, 0.9$.)

(i) $\phi(x) = \sin x$

$||f(a^o|| = 0.003\ 545$

$a^* = (-0.011\ 342,\ 1.123\ 046,\ 0.299\ 217)^T$

$||f(a^*)|| = 0.003\ 230, \ m = 3 \ , \ N = 2 \ , \ N_e = 3 \ .$

(ii) $\phi(x) = \sqrt{x}$

$||f(a^o)|| = 0.008\ 033$

$a^* = (0.175\ 838 \ , \ 1.716\ 68 \ , \ 0.913\ 932)^T$

$||f(a^*)|| = 0.007\ 484 \ , \ m = 3 \ , \ N = 3 \ , \ N_e = 4$

(iii) $\phi(x) = 7 \sin(15x)$

(In this case, a^o gave a pole, and was reset to $(1,1,1)^T$.)

$||f(a^o)|| = 4.439\ 981 \ , \ m = 5 \ ,$

$a^* = (4.974\ 616 \ , \ -1.246\ 795 \ , \ 7.023\ 808)^T \ ,$

$$||f(a^*)|| = 4.337\ 718\ ,\ m = 5\ ,\ N = 18\ ,\ N_e = 25\ .$$

Example 4

$$f(x,a) = a_1 + \sum_{i=1}^{3} a_{2i} \cos(a_{2i+1}\, x) - \frac{1}{\pi(1+x^2)}, \quad X = [0,2]$$

$$a^o = (0,0,1,0,4,0,6)^T, \quad ||f(a^o)|| = 0.352\ 416\ ,\ m = 0\ ,$$

$$a^* = (0.178\ 250\ ,\ 0.121\ 330\ ,\ 1.627\ 427\ ,\ 0.017\ 075\ ,\ 3.781\ 941\ ,\ 0.001\ 537\ ,\ 6.327\ 107)^T, \quad ||f(a^*)|| = 0.000\ 122\ ,\ m = 7\ ,\ N = 50\ ,\ N_e = 115\ .$$

In this example, only the last 6 iterations used H defined by (3.3).

REFERENCES

1. Anselone, P. M. and W. Krabs, Approximate solution of weakly singular integral equations, Journal of Integral Equations (to appear).

2. Clarke, F. H., A new approach to Lagrange multipliers, Mathematics of Operations Research 1 (1976), 165-174.

3. Collatz, L. and W. Krabs, Approximationstheorie, Stuttgart, Teubner - Verlag 1973.

4. Fletcher, R., A model algorithm for composite NDO problems, Mathematical Programming Studies (to appear).

5. Fletcher, R., Practical Methods of Optimization, Vol II Constrained Optimization, Chichester, Wiley 1981.

6. Glashoff, K. and R. Schultz, Uber die genaue Berechnung von besten L^1 - Approximierenden, J. Approx. Th. 25 (1979), 280-293.

7. Hettich, R. and W. van Honstede, On quadratically convergent methods for semi-infinite programming, in Semi-Infinite Programming, Proc., ed. R. Hettich. Berlin, Springer-Verlag 1979.

8. van Honstede, W., An approximation method for semi-infinite programming, in Semi-Infinite Programming, Proc., ed R. Hettich, Berlin, Springer-Verlag 1979.

9. Krabs, W., One-sided L_1 approximation as a problem of semi-infinite programming, in Semi-Infinite Programming, Proc., ed R. Hettich. Berlin, Springer-Verlag 1979.

10. Marsaglia, G., One-sided approximations by linear combinations of functions, in Approximation Theory, ed A. Talbot. London, Academic Press 1970.

11. Watson, G. A., An algorithm for linear L_1 approximation of continuous functions, I.M.A.J. Num Anal. (to appear).

12. Watson, G. A., Globally convergent methods for semi-infinite programming, Dundee University Mathematics Department Report NA/45 (1981).

Department of Mathematics
University of Dundee
Dundee DD1 4HN
Scotland

APPLICATIONS OF THE GENERALIZED QD ALGORITHM. *

Guido Claessens
Luc Wuytack

The generalized QD algorithm can be used to construct continued fractions whose convergents are multipoint Padé approximants. A convergence result for multipoint Padé approximants, having the same degree in the denominator, is proved. This result generalizes the Koenig-Hadamard theorem for the classical case of Padé approximation.

It is shown how this convergence result implies some convergence properties for the rows and columns in the extended QD-table. These properties are similar to the convergence properties of Rutishauser's classical QD-table.

The convergence properties of the generalized QD-table can be used to obtain approximations for all the zeros and poles of a meromorphic function, whose derivatives are given at a finite number of points.

1. INTRODUCTION

In this paper we will concentrate on certain types of local rational approximations for a function f, given in the form of a power series or as a Newton interpolating series.

First we will survey some of the results from the theory of Padé approximation (section 2). Convergence properties of columns in the Padé table will be given. It will also be indicated how Padé approximants can be seen as convergents of certain type of continued fractions. The coefficients in these continued fractions can be computed by using Rutishauser's qd-algorithm. In theorem 3 it is stated that the columns in the qd-table have some convergence properties that can be used to find the poles of a meromorphic function.

This theory is now extended to the case of multipoint Padé approximation. Assume a function f and its derivatives are given at a finite

---

* Work supported in part by the NFWO (Belgium).

number of points. Then multipoint Padé approximants can be defined and in theorem 4 (section 3) an explicit expression is given for their numerator and denominator.

In section 4 the convergence of multipoint Padé approximants is considered and a generalization of the Koenig-Hadamard theorem is obtained. In section 5 it is shown how multipoint Padé approximants can be related to certain types of continued fractions, whose coefficients can be computed by using the generalized qd-algorithm. In theorem 7 it is shown how the convergence of the extended (generalized) qd-table is related to the location of the poles of a meromorphic function.

In section 6 some applications are mentioned. These applications include : finding zeros and poles of a function whose derivatives are given at a finite number of points. In section 7 some numerical examples are given to illustrate the convergence properties of the generalized qd-algorithm.

2. THE PADE APPROXIMATION PROBLEM AND THE QD ALGORITHM.

Let f be a given power series or $f(x) = c_0+c_1.x+c_2.x^2+ \ldots$ with $c_0 \neq o$. Let R_n^m denote the class of (ordinary) rational functions $r = \frac{p}{q}$, where p and q are polynomials of degree at most m and n respectively such that $\frac{p}{q}$ is irreducible. The Padé approximation problem for f of order (m,n) is to find an element $r = \frac{p}{q}$ in R_n^m satisfying $f(x).q(x)-p(x) = O(x^{m+n+1+j})$, where j is an integer which is as large as possible. This problem has a unique solution for all values of m and n, denoted by $r_{m,n} = \frac{p_{m,n}}{q_{m,n}}$. The elements $r_{m,n}$ can be ordered in a two-dimensional array, called the Padé table, as follows :

$$
\begin{array}{ccccc}
r_{00} & r_{01} & r_{02} & r_{03} & \cdots \\
r_{10} & r_{11} & r_{12} & r_{13} & \cdots \\
r_{20} & r_{21} & r_{22} & r_{23} & \cdots \\
r_{30} & r_{31} & r_{32} & r_{33} & \cdots \\
\cdot\cdot & \cdot\cdot & \cdot\cdot & \cdot\cdot &
\end{array}
$$

This table has a "block structure" which means that equal elements appear in square blocks. A Padé table is called normal if all its elements are different from each other. The elements in a column in this table have the following convergence property (see [10], p. 118; [11], p. 235).

THEOREM 1. Let f be meromorphic in the disc $D_R = \{x : |x| < R\}$, having $\{s_1, s_2, \ldots, s_n\}$ as poles in D_R, with $|s_1| \leq |s_2| \leq \ldots \leq |s_n| \leq c.R < R$ for some constant c. Let $h(x) = (x-s_1).(x-s_2) \ldots (x-s_n)$ and $s_i^{(m)}$ be the zeros of $q_{m,n}$ for $i = 1,2, \ldots, n$. Then $s_i^{(m)} = s_i + O(c^m)$ for $i = 1,2, \ldots, n$ and $q_{m,n} = h + O(c^m)$ as $m \to \infty$.

The elements on a "staircase" $S_k = \{r_{k,0}, r_{k+1,0}, r_{k+1,1}, r_{k+2,1}, r_{k+2,2}, \ldots\}$ in the Padé table can be seen as convergents of certain types of continued fractions (see [13], p. 258).

THEOREM 2. If the Padé table for f is normal, then there exist continued fractions of the form

$$\sum_{i=0}^{k} c_i.x^i + \frac{c_{k+1}.x^{k+1}|}{|1} - \frac{q_1^{k+1}.x|}{|1} - \frac{e_1^{k+1}.x|}{|1} - \frac{q_2^{k+1}.x|}{|1} - \frac{e_2^{k+1}.x|}{|1}$$

$$- \frac{q_3^{k+1}.x|}{|1} - \frac{e_3^{k+1}.x|}{|1} - \ldots \qquad (1)$$

with $q_i^{k+1} \neq 0$, $e_i^{k+1} \neq 0$ whose convergents are the elements of S_k for $k \geq 0$.

The coefficients in the continued fraction (1) can be computed by using Rutishauser's qd algorithm (see [14]).

Let $e_0^m = 0$ and $q_1^m = \dfrac{c_{m+1}}{c_m}$ for $m \geq 1$.

$$\left.\begin{aligned} \text{Then}\quad e_n^m &= e_{n-1}^{m+1} + q_n^{m+1} - q_n^m \\ q_{n+1}^m &= q_n^{m+1}.\frac{e_n^{m+1}}{e_n^m} \end{aligned}\right\} \text{ for } m \geq 1 \text{ and } n \geq 1. \qquad (2)$$

The elements e_n^m, q_n^m can be ordered in a scheme (the qd-table) as follows

$$
\begin{array}{ccccccc}
 & q_1^{0} & & q_2^{-1} & & q_3^{-2} & \\
e_0^{1} & & e_1^{0} & & e_2^{-1} & & \cdots \\
 & q_1^{1} & & q_2^{0} & & q_3^{-1} & \\
e_0^{2} & & e_1^{1} & & e_2^{0} & & \cdots \\
 & q_1^{2} & & q_2^{1} & & q_3^{0} & \\
e_0^{3} & & e_1^{2} & & e_2^{1} & & \cdots \\
 & q_1^{3} & & q_2^{2} & & q_3^{1} & \\
 -- & & -- & & -- & &
\end{array}
$$

The same recurrence relations as (2) also hold for the elements above the diagonal, with $q_n^{-n+1} = 0$ for $n > 1$ and $q_1^0 = - \frac{c_1}{c_0}$.

Based on theorem 1 and 2 it is possible to prove the following convergence property for the columns in the qd-table (see [12], [8] p. 612).

<u>THEOREM 3</u>. Let f be meromorphic in the disc $D_R = \{x : |x| < R\}$, having $\{s_1, s_2, \ldots, s_n\}$ as poles in D_R, with $|s_1| \leqslant |s_2| \leqslant |s_3| \leqslant \ldots$. Then the qd-table has the properties :

(a) if $|s_{m-1}| < |s_m| < |s_{m+1}|$ then $\lim\limits_{n\to\infty} q_m^n = \frac{1}{s_m}$.

(b) if $|s_m| < |s_{m+1}|$ then $\lim\limits_{n\to\infty} e_m^n = 0$.

Similar results hold if certain poles have the same modules.

This theorem has several applications, as described by P. Henrici in e.g. [8] and [9]. It implies that the qd-algorithm can be used to find the zeros of polynomials and entire functions. It also can be used to find the zeros and poles of a meromorphic functions, whose power series is known at a certain point or, equivalently, whose derivatives are known at a certain point.

<u>3. MULTIPOINT PADE APPROXIMATION.</u>

Let $x_1, x_2, \ldots, x_k$ be given and assume $f_i^{(j)}$ is the j-th derivative at the point x_i of a function f, for $j \geqslant 0$. Let $y_{k.j+i-1} = x_i$ for $j \geqslant 0$

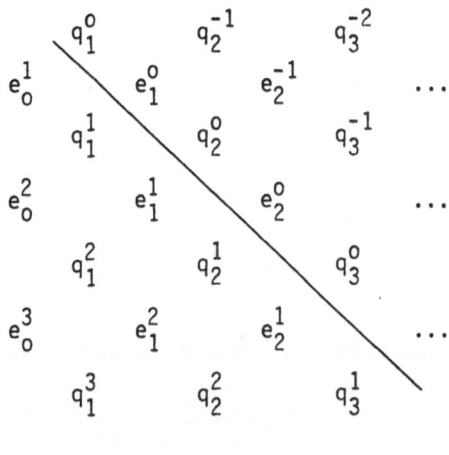

and $i = 1,2,\ldots,k$, and $\omega_i(x) = (x-y_{i-1}).\omega_{i-1}(x)$ for $i \geq 1$, with $\omega_0(x) = 1$. It is then possible to construct the Newton interpolating series for the given data

$$f(x) = c_0 + \sum_{i=1}^{\infty} c_i.\omega_i(x),$$

where the c_i are divided differences with confluent arguments (see [18], p. 53, [6], p. 66).

The multipoint Padé approximation problem is to find an element $r = \frac{p}{q}$ in R_n^m such that

$$f(x).q(x) - p(x) = O(\omega_j(x)) \tag{3}$$

where j is an integer which is as large as possible. It is known that there exists a unique solution for this problem (see [4], [19]). This solution will be denoted by $r_{m,n} = \frac{p_{m,n}}{q_{m,n}}$, normalized in a suitable way, and ordered in a table like the Padé table. This table will be called the MP-table.

Let $p(x) = \sum_{i=0}^{m} a_i.\omega_i(x)$ and $q(x) = \sum_{i=0}^{n} b_i.\omega_i(x)$ then (3) implies that a_i, b_i must satisfy the following system of linear equations (see [5], [19]) :

$$\sum_{j=0}^{k} b_j.f_{jk} = a_k \quad \text{for} \quad k = 0,1,\ldots,m \tag{4a}$$

$$\sum_{j=0}^{k} b_j.f_{jk} = 0 \quad \text{for} \quad k = m+1,\ m+2,\ \ldots,\ m+n \tag{4b}$$

Here f_{jk} for $k \geq j$ denotes the divided difference of order k-j of f at the points $y_j, y_{j+1},\ldots,y_k$ and $f_{jk} = 0$ if $k > j$.

<u>THEOREM 4</u>. If the rank of (4b) is maximal then a solution of (3) can be written as follows [5][19] :

$$p_{m,n}(x) = \begin{vmatrix} f_{n,m+1} & f_{n,m+2} & \cdots & f_{n,m+n} & \omega_n(x).F_{nm}(x) \\ f_{n-1,m+1} & f_{n-1,m+2} & \cdots & f_{n-1,m+n} & \omega_{n-1}(x).F_{n-1,m}(x) \\ \cdots & \cdots & \cdots & \cdots & \cdots \\ f_{o,m+1} & f_{o,m+2} & \cdots & f_{o,m+n} & \omega_o(x).F_{o,m}(x) \end{vmatrix}$$

$$q_{m,n}(x) = \begin{vmatrix} f_{n,m+1} & f_{n,m+2} & \cdots & f_{n,m+n} & \omega_n(x) \\ f_{n-1,m+1} & f_{n-1,m+2} & \cdots & f_{n-1,m+n} & \omega_{n-1}(x) \\ \cdots & \cdots & \cdots & \cdots & \cdots \\ f_{o,m+1} & f_{o,m+2} & \cdots & f_{o,m+n} & \omega_o(x) \end{vmatrix}$$

where $F_{i,m}(x) = \sum_{j=i}^{m} f_{ij}\,\omega_{ij}(x)$ for $m \geqslant i$, with the convention that $F_{i,m}(x) \equiv 0$

if $i > m$ and $\omega_{ij}(x) = (x-y_{j-1}).\omega_{i,j-1}(x)$, with $\omega_{ii}(x) = 1$.

4. CONVERGENCE OF MULTIPOINT PADE APPROXIMANTS.

In order to prove a similar result as in theorem 1 we need some notations. Let $E_R = \{x : |(x-x_1)(x-x_2) \ldots (x-x_k)| < R\}$. By p_n we denote the polynomial of degree at most n that interpolates f at the points $\{y_0,y_1,\ldots,y_n\}$. Concerning the convergence of p_n to f as $n \to \infty$, Walsh [18, p. 61] proved the following result.

LEMMA 1. Let f be analytic in E_R, then $\lim_{n\to\infty} p_n(x) = f(x)$ for all x in E_R. The convergence is uniform in any closed set lying interior to E_R. More precisely, we have $|f(x) - p_n(x)| \leqslant M.(\frac{R'}{R})^n$ for x in $E_{R'}$, with $R' < R$ and where M is independent of n and x, but not from R and R'.

Using this result we can prove a convergence theorem for the multipoint Padé approximants of a meromorphic function. This theorem is a generalization of the Koenig-Hadamard theorem for Padé approximants (theorem 1) and can also be seen as a generalization of the theorem of the Montessus de Ballore [12].

First we introduce generalized Vandermonde determinants $V(z_1,z_2,\dots,z_n)$ and prove some auxiliary results.

Let $z_1,z_2,\dots,z_n$ be complex numbers, which are different from each other. The generalized Vandermonde determinant $V(z_1,z_2,\dots,z_n)$ be defined as follows :

$$V(z_1,z_2,\dots,z_n) = \begin{vmatrix} 1 & 1 & \dots & 1 \\ \omega_1(z_1) & \omega_1(z_2) & \dots & \omega_1(z_n) \\ \omega_2(z_1) & \omega_2(z_2) & \dots & \omega_2(z_n) \\ \vdots & \vdots & & \vdots \\ \omega_{n-1}(z_1) & \omega_{n-1}(z_2) & \dots & \omega_{n-1}(z_n) \end{vmatrix}$$

<u>LEMMA 2.</u> $V(z_1,z_2,\dots,z_n) = \prod_{j<i} (z_i-z_j)$ with $1 \leq i,j \leq n$.

<u>Proof</u>. Let $W(z) = V(z_1,z_2,\dots,z_{n-1},z)$, then $W(z)$ is a polynomial of degree $n-1$. This polynomial vanishes in the points $z_1,z_2,\dots,z_{n-1}$, since replacing z by one of these values causes two identical rows in the determinant. Consequently

$$V(z_1,z_2,\dots,z_{n-1},z)= C.(z-z_1)(z-z_2)\ \dots\ (z-z_{n-1}).$$

If we expand the determinant by minors of the last row, then we see that the coefficient of $\omega_{n-1}(z)$ is $V(z_1,z_2,\dots,z_{n-1})$. Thus, we have

$$V(z_1,z_2,\dots,z_{n-1},z) = V(z_1,z_2,\dots,z_{n-1})(z-z_1)(z-z_2)\dots(z-z_{n-1}) \quad (5)$$

Hence we have the recursion formula

$$V(z_1,z_2,\dots,z_n) = V(z_1,z_2,\dots,z_{n-1})(z_n-z_1)(z_n-z_2)\dots(z_n-z_{n-1}).$$

And since $V(z_1,z_2) = z_2-z_1$, we get

$$V(z_1,z_2,\dots,z_n) = \prod_{j<i} (z_i-z_j),\ 1 \leq i,j \leq n.$$

Another proof of this result is given by D. Warner [19, p. 44].
The above proof also shows that

$$\frac{V(z_1,z_2,\dots,z_{n-1},z)}{V(z_1,z_2,\dots,z_{n-1})} = \prod_{i=1}^{n-1} (z-z_i).$$

We remark that this relation continues to hold even if some of the elements $z_1, z_2, \ldots, z_{n-1}$ are equal. This is very easily shown using L'Hospital's rule for the indeterminate form on the left.

Let $f = \frac{g}{h}$ where $g(x) = \sum_{i=0}^{\infty} g_i . \omega_i(x)$ and $h(x) = \sum_{i=0}^{\infty} h_i . \omega_i(x)$, with $h_{ii} \neq 0$ for $i \geqslant 0$. Using theorem 4 it is possible to prove (see [5], [19]) the following result, in case the rank of the corresponding system (4.b) is maximal.

LEMMA 3. Let $f = \frac{g}{h}$ then $r_{m,n} = \frac{p_{m,n}}{q_{m,n}}$ can be expressed as follows :

$$p_{m,n}(x) = \begin{vmatrix} h_{oo} & h_{o1} & \cdots & h_{o,m+n} & -\omega_o(x) \\ o & h_{11} & \cdots & h_{1,m+n} & -\omega_1(x) \\ o & o & \cdots & h_{2,m+n} & -\omega_2(x) \\ \vdots & \vdots & & \vdots & \vdots \\ o & g_{11} & \cdots & g_{1,m+n} & o \\ g_{oo} & g_{o1} & \cdots & g_{o,m+n} & o \end{vmatrix} \tag{6}$$

and

$$q_{m,n}(x) = \begin{vmatrix} h_{oo} & h_{o1} & \cdots & h_{o,m+n} & o \\ o & h_{11} & \cdots & h_{1,m+n} & o \\ o & o & & h_{2,m+n} & o \\ \vdots & \vdots & & \vdots & \vdots \\ o & g_{11} & \cdots & g_{1,m+n} & \omega_1(x) \\ g_{oo} & g_{o1} & \cdots & g_{o,m+n} & \omega_o(x) \end{vmatrix} \tag{7}$$

The expressions (6) and (7) can also be written as

$$p_{m,n}^{(x)} = g_{oo} . \delta[(h)_m, (g)_n]_1 . \omega_o(x) + \ldots + (-1)^n . \delta[(h)_m, (g)_{n+1}]_o . \omega_m(x) \tag{6'}$$

and $$q_{m,n}^{(x)} = h_{oo} . \delta[(h)_m, (g)_n]_1 . \omega_o(x) + \ldots + (-1)^n . \delta[(h)_{m+1}, (g)_n]_o . \omega_n(x) \tag{7'}$$

where $\delta[(h)_m, (g)_n]_i$ are generalized bigradients defined as follows :

$$\delta[(h)_m,(g)_n]_i = \begin{vmatrix} h_{ii} & h_{i,i+1} & h_{i,i+2} & \cdots & h_{i,i+m+n-1} \\ o & h_{i+1,i+1} & h_{i+1,i+2} & \cdots & h_{i+1,i+m+n-1} \\ o & o & h_{i+2,i+2} & \cdots & h_{i+2,i+m+n-1} \\ \vdots & \vdots & \vdots & & \vdots \\ o & o & \vdots & & h_{i+m-1,i+m+n-1} \\ o & o & \vdots & & g_{i+n-1,i+m+n-1} \\ \vdots & \vdots & \vdots & & \vdots \\ o & g_{i+1,i+1} & g_{i+1,i+2} & \cdots & g_{i+1,i+m+n-1} \\ g_{ii} & g_{i,i+1} & g_{i,i+2} & \cdots & g_{i,i+m+n-1} \end{vmatrix}$$

Before proving the generalization of the Montessus de Ballore's theorem, we note that there exists a more general theorem, due to Saff [15]. His theorem concerns interpolation sequences, while ours is only valid for interpolation series, which are a special case of interpolation sequences. Our proof depends mainly upon the above explicit bigradiental representation of the multipoint Padé approximants, and is a generalization of Householder's proof [10] for the Padé case. We assume that $r_{m,n}$ is normalized such that $q_{m,n}$ is monic.

<u>THEOREM 5</u>. Let f be meromorphic in E_R and analytic in the points $\{x_1,x_2,\ldots,x_k\}$. Let $\{s_1,s_2,\ldots,s_n\}$ be the poles of f in E_R, with $o < R_1 \leq R_2 \leq \ldots \leq R_n \leq c.R < R$ for some constant c and with $R_i = |(s_i-x_1)(s_i-x_2)\ \ldots\ (s_i-x_k)|$. Let $h(x) = (x-s_1).(x-s_2)\ \ldots\ (x-s_n)$ and $s_i^{(m)}$ be the zeros of $q_{m,n}$ for $i = 1,2,\ldots,n$. Then $s_i^{(m)} = s_i + O(c^m)$ for $i = 1,2,\ldots,n$ and $q_{m,n} = h + O(c^m)$ as $m \to \infty$.

<u>Proof</u>. Consider the formal Newton series

$$f(z) = \sum_{i=0}^{\infty} f_{oi}.\omega_i(z).$$

Let $g(z) = f(z).h(z)$, then $g(z)$ is analytic in E_R. And moreover $g(s_i) \neq o$ for $i = 1,2,\ldots,n$. For every $i \geq o$, we can formally determine $\sum_{j=i}^{\infty} g_{ij}\ \omega_{ij}(z)$. By Lemma 1 we know then that $\sum_{j=i}^{\infty} g_{ij}\ \omega_{ij}(z)$ represents $g(z)$

in E_R.

Let us now write g(z) as follows

$$g(z) = G_{i,n-1}(z) + \omega_{in}(z).d_{in}(z)$$

with $G_{i,n-1}(z) = \sum_{j=i}^{n-1} g_{ij}.\omega_{ij}(z)$ and $d_{in}(z) = \sum_{j=n}^{\infty} g_{ij}.\omega_{ij}(z)$.

Using Lemma 1 we get, for m sufficiently large and $R_i \leqslant c.R < R$, for some constant c and i = 1,2,...,n, that

$$G_{j,m+n}(s_i) = g(s_i) + O(c^m) \text{ for } j = o,1,\dots,n. \tag{8}$$

Let us now first consider the case where $s_1, s_2, \dots, s_n$ are all distinct. Using (7') and the normalization condition for $q_{m,n}$ we get if $\delta[(h)_{m+1}, (g)_n]_o \neq o$,

$$q_{m,n}(z) = \frac{(-1)^n.Q_{m,n}(z)}{\delta[(h)_{m+1},(g)_n]_o} \tag{9}$$

where

$$Q_{m,n}(z)=\begin{vmatrix} h_{oo} & h_{o1} & \cdots & h_{o,n-1} & 1 & o & \cdots & o & o \\ o & h_{11} & \cdots & h_{1,n-1} & h_{1,n} & 1 & \cdots & o & o \\ \vdots & \vdots & & \vdots & \vdots & \vdots & & \vdots & \vdots \\ & & & & & & & 1 & o \\ o & o & & o & g_{n,n} & g_{n,n+1} & \cdots & g_{n,m+n} & \omega_n(z) \\ \vdots & \vdots & & \vdots & \vdots & \vdots & & \vdots & \vdots \\ o & g_{11} & \cdots & g_{1,n-1} & g_{1,n} & g_{1,n+1} & \cdots & g_{1,m+n} & \omega_1(z) \\ g_{oo} & g_{o1} & \cdots & g_{o,n-1} & g_{o,n} & g_{o,n+1} & \cdots & g_{o,m+n} & \omega_o(z) \end{vmatrix}$$

To evaluate $\delta[(h)_{m+1},(g)_n]$ we consider $P_1 = \delta[(h)_{m+1},(g)_n.V(s_1,s_2,\dots,s_n)$.

Using the fact that $h(s_i) = o$ for $i = 1,2,\dots,n$ we get

$$P_1 = (-1)^{n(m+1)} \cdot \begin{vmatrix} \omega_{n-1}(s_1).G_{n-1,m+n}(s_1) & \cdots & \omega_{n-1}(s_n).G_{n-1,m+n}(s_n) \\ \omega_{n-2}(s_1).G_{n-2,m+n}(s_1) & \cdots & \omega_{n-2}(s_n).G_{n-2,m+n}(s_n) \\ \cdots & \cdots & \cdots \\ \omega_0(s_1).G_{0,m+n}(s_1) & \cdots & \omega_0(s_n).G_{0,m+n}(s_n) \end{vmatrix}$$

The same reasoning can be used to evaluate $Q_{m,n}(z)$. Let $P_2 = Q_{m,n}(z).V(s_1,s_2,\ldots,s_n)$ then we get

$$P_2 = (-1)^{n.(m+1)} \cdot \begin{vmatrix} \omega_n(s_1).G_{n,m+n}(s_1) & \cdots & \omega_n(s_n).G_{n,m+n}(s_n) & \omega_n(z) \\ \omega_{n-1}(s_1).G_{n-1,m+n}(s_1) & \cdots & \omega_{n-1}(s_n).G_{n-1,m+n}(s_n) & \omega_{n-1}(z) \\ \cdots & \cdots & \cdots & \cdots \\ \omega_0(s_1).G_{0,m+n}(s_1) & \cdots & \omega_0(s_n).G_{0,m+n}(s_n) & \omega_0(z) \end{vmatrix}$$

Hence, using (8), we get

$$P_1 = (-1)^{n.(m+1)+\frac{(n-1).n}{2}} \cdot \left[\prod_{i=1}^{n} g(s_i)\right].V(s_1,s_2,\ldots,s_n)+0(c^m).$$

and

$$P_2 = (-1)^{n.(m+1)+\frac{(n+1).n}{2}} \cdot \left[\prod_{i=1}^{n} g(s_i)\right].V(s_1,s_2,\ldots,s_n,z)+0(c^m).$$

Consequently $\delta[(h)_{m+1},(g)_n]_0 = (-1)^{n.(m+\frac{n+1}{2})} \cdot \left[\prod_{i=1}^{n} g(s_i)\right] +0(c^m)$,

which implies that $\delta[(h)_{m+1},(g)_n]_0 \neq 0$ for m sufficiently large. Since $q_{m,n}(z) = (-1)^n . \frac{P_2}{P_1}$ we get, using (9) and (5),

$$q_{m,n}(z) = \prod_{i=1}^{n} (z-s_i) + O(c^m) \quad \text{for } m \to \infty .$$

This proves the theorem in case of distinct poles. In view of the remark following Lemma 2 and the form of P_1,P_2 we can derive a similar result in case some of the poles are equal.

5. THE GENERALIZED QD-ALGORITHM.

Let f be defined as in the preceding section or $f(x) = = c_0 + \sum_{i=1}^{\infty} c_i.\omega_i(x)$. Consider the elements $T_k = \{r_{k,0}, r_{k+1,0}, r_{k+1,1}, r_{k+2,1}, r_{k+2,2}, \ldots\}$ in the MP-table for $k \geq 0$. In the normal case (i.e. all elements in the MP-table are different), the elements of T_k can be considered as convergents of continued fractions of a certain form.

THEOREM 6. If the MP-table is normal, then there exists continued fractions of the form

$$\sum_{i=0}^{k} c_i.\omega_i(x) + \frac{c_{k+1}.\omega_{k+1}(x)|}{|1} - \frac{q_1^{k+1}.(x-y_{k+1})|}{|1+q_1^{k+1}.(y_0-y_{k+1})} - \frac{e_1^{k+1}.(x-y_{k+2})|}{|1+e_1^{k+1}.(y_0-y_{k+2})}$$

$$- \frac{q_2^{k+1}.(x-y_{k+3})|}{|1+q_2^{k+1}.(y_0-y_{k+3})} - \frac{e_2^{k+1}.(x-y_{k+4})|}{|1+e_2^{k+1}.(y_0-y_{k+4})} - \ldots \qquad (10)$$

with $c_{k+1} \neq 0$, $q_j^{k+1} \neq 0$, $e_j^{k+1} \neq 0$ for $j \geq 0$, whose convergents are the elements of T_k (normalized such that $q_{m,n}(y_0) = 1$), for $k \geq 0$.

An algorithm, called the generalized qd-algorithm, that can be used to compute the coefficients in (10) is given in [3]. This algorithm computes the elements in a qd-table as follows. As before f_{jk} denotes the divided difference of order k-j at $y_j, y_{j+1}, \ldots, y_k$.

$$\text{Let } e_0^m = 0 \text{ and } e_1^m = \frac{f_{0,m+1}}{f_{1,m+1}} \text{ for } m \geq 1. \qquad (11)$$

$$\text{Then} \left\{ \begin{array}{l} e_n^m = \dfrac{q_n^{m+1} - q_n^m + e_{n-1}^{m+1}.[1-q_n^{m+1}.(y_0-y_{m+2.n-1})]}{1+q_n^m.(y_0-y_{m+2.n-1})} \\[2em] q_{n+1}^m = \dfrac{e_n^{m+1}.q_n^{m+1}.[1+e_n^m.(y_0-y_{m+2.n-1})]}{e_n^m.[1+q_n^{m+1}.(y_0-y_{m+2n-1})]+e_n^{m+1}.(e_n^m-q_n^{m+1}).(y_0-y_{m+2n+1})} \end{array} \right.$$

for $m \geq 1$ and $n \geq 1$. (12)

These relations were proved using recurrence relations between generalized Hankel determinants. These determinants $H_k^{l,m}$ are defined as follows for $l,m \geq o$ and $o < k \leq l+1$:

$$H_k^{l,m} = \begin{vmatrix} f_{l,m} & f_{l,m+1} & \cdots & f_{l,m+k-1} \\ f_{l-1,m} & f_{l-1,m+1} & \cdots & f_{l-1,m+k-1} \\ \vdots & \vdots & & \vdots \\ f_{l-k+1,m} & f_{l-k+1,m+1} & \cdots & f_{l-k+1,m+k-1} \end{vmatrix}$$

and $H_o^{l,m} = 1$. In [3], the following result is proved

LEMMA 4. For $n \geq 1$ and $m \geq o$ we have

$$q_n^{m+1} = \frac{H_n^{n-1,m+n+1}}{H_{n-1}^{n-2,m+n}} \cdot \frac{H_{n-1}^{n-1,m+n}}{H_n^{n,m+n+1}}$$

$$e_n^{m+1} = \frac{H_{n+1}^{n,m+n+1}}{H_n^{n-1,m+n}} \cdot \frac{H_{n-1}^{n-1,m+n+1}}{H_n^{n,m+n+2}}$$

Using this property it can be shown (see [3]) that the relations (12) also hold between the elements above the diagonal in the extended qd-table. The elements in the first rows must be defined as follows, with $g = \frac{1}{f}$:

$$q_n^{1-n} = o \text{ for } n \geq 2; \; q_1^0 = -\frac{g_{01}}{g_{00}} \; ; \; e_n^{1-n} = \frac{g_{0,n+1}}{g_{1,n+1}} \text{ for } n \geq 1. \tag{13}$$

The quotient of certain generalized Hankel determinants of f is now related to the poles of f as follows.

THEOREM 6. Let the assumptions of theorem 5 hold, then

$$\frac{H_n^{n,m+1}}{H_n^{n-1,m+1}} = \prod_{i=1}^{n} (s_i - y_o) + O(c^m), \text{ as } m \to \infty .$$

Proof. Let $q_{m,n}(x) = \sum_{i=0}^{n} b_i \cdot \omega_i(x)$ then theorem 5 implies that

$$b_o = q_{m,n}(y_o) = \prod_{i=1}^{n} (y_o - s_i) + O(c^m).$$

As a result of theorem 1 and the normalization condition for $r_{m,n}$ we get $b_o = (-1)^n . \frac{H_n^{n,m+1}}{H_n^{n-1,m+1}}$. Combining these relations gives the required result.

It is now possible to prove a convergence property for the columns in the qd-table in case f is a meromorphic function whose poles satisfy certain relations.

THEOREM 7. Let f be meromorphic in E_R and analytic in $\{x_1, x_2, \ldots, x_k\}$. Let $\{s_1, s_2, \ldots, s_n\}$ be the poles of f in E_R and $R_i = |(s_i - x_1)(s_i - x_2) \ldots$ $\ldots (s_i - x_k)|$. Then the extended (generalized) qd-table has the properties :

(a) if $R_{m-1} < R_m < R_{m+1}$, with $R_o = o$ and $R_{n+1} = R$, then

$$\lim_{n\to\infty} q_m^n = \frac{1}{s_m - y_o},$$

(b) if $R_m < R_{m+1}$, then

$$\lim_{n\to\infty} e_m^n = o.$$

Proof. Using theorem 6 we get, since $R_{m-1} < R_m$ and $R_m < R_{m+1}$,

$$\lim_{n\to\infty} \frac{H_{m-1}^{m-1,n+m-1}}{H_{m-1}^{m-2,n+m-1}} = \prod_{i+1}^{m-1} (s_i - y_o)$$

$$\lim_{n\to\infty} \frac{H_m^{m,n+m}}{H_m^{m-1,n+m}} = \prod_{i=1}^{m} (s_i - y_o)$$

Hence, using lemma 4, we get $\lim_{n\to\infty} q_m^n = \frac{1}{s_m - y_o}$. Applying the forward recurrence relation for continued fractions to (10) gives :

$$q_{n+m,m} = [1 + e_m^n . (y_o - y_{n+2m-1})] . q_{n+m-1,m} - e_n^m . (x - y_{n+2m-1}) . q_{n+m-1,m-1}$$

Taking the limit for $n \to \infty$ and using theorem 5 we get $\lim_{n\to\infty} e_m^n = o$.

Remark that from the proof of the above theorem it follows that the rate of convergence of q_m^n towards $\frac{1}{s_m - y_0}$, depends on the ratios $\frac{R_{m-1}}{R_m}$ and $\frac{R_m}{R_{m+1}}$. Hence the choice of $x_1, x_2, \ldots, x_k$ can strongly influence this convergence behaviour.

In case $R_{m-1} = R_m$ or $R_m = R_{m+1}$ then polynomials can be constructed (by using values from the qd-table), whose zeros are the poles of f. Theorems of this kind are given and proved in [5].

We also remark that, due to some symmetry properties of the MP-tables for f and $\frac{1}{f}$, it is possible to prove convergence properties (as in theorem 5) for the numerators in a row of the MP-table of a meromorphic function. This result then implies the following converging properties for the rows in the generalized qd-table. We refer to [5] for more details.

<u>THEOREM</u> 8. Let f be a meromorphic function in E_R and analytic in $\{x_1, x_2, \ldots, x_k\}$. Let $\{z_1, z_2, \ldots, z_k\}$ be the zeros of f in E_R and $R_i' = |(z_i - x_1)(z_i - x_2) \ldots (z_i - x_k)|$. Then the extended (generalized) qd-table has the properties :

(a) if $R'_{m-1} < R'_m < R'_{m+1}$, with $R'_0 = o$ and $R'_{\ell+1} = R$, then

$$\lim_{n\to\infty} e_{n+m-1}^{1-n} = \frac{1}{z_m - y_0} .$$

(b) if $R'_m < R'_{m+1}$, then

$$\lim_{n\to\infty} q_{n+m-1}^{1-n} = o.$$

6. SOME APPLICATIONS.

In general we remark that the preceding theory can be applied in the cases where the derivatives of a function f are known at several points. This situation occurs e.g. in certain problems of physics (see [1], [2], [7]) and in the reduction of linear control systems [16], [17]. The given information about f can then be used to find its zeros and poles. We will now describe some of these applications in more detail.

6.a. Zeros of polynomials.

Suppose we want to determine the zeros of a Newton interpolating polynomial of degree n

$$p_n(x) = \sum_{i=0}^{n} f_{oi}\ \omega_i(x)$$

These zeros can be found as the poles of the rational function

$$f(z) = \frac{1}{p_n(z)} .$$

Since in this case we know the reciprocal function of f(z), we can construct the first two rows of the generalized qd-table. We get

$$q_k^{-k+1} = \begin{cases} -\dfrac{f_{o1}}{f_{oo}} & \text{for } k = 1 \\ o \text{ for } k > 1 \end{cases}$$

$$e_k^{-k+1} = \begin{cases} \dfrac{f_{o,k+1}}{f_{1,k+1}} & \text{for } k = 1,2,\dots,n-1 \\ o \text{ for } k \geqslant n. \end{cases}$$

Moreover, since $r_{m,k} = f$ for $k \geqslant n$, we know that $e_n^k = o$ for $k \geqslant 1-n$. The qd-table can now be generated row by row using the formulas (12). If $R_{m-1} < R_m < R_{m+1}$ then theorem 7 implies that q_m^k converges to $\frac{1}{s_m - y_o}$, as $k \to \infty$ where s_m is a zero of p_n. This means that under the given condition the q-columns in the qd-table will convergence to the inverses of the (shifted) zeros of p_n.

Remark that different zeros, which have the same absolute value, can be obtained as the inverses of limits of q-columns, provided the choice of $x_1, x_2, \dots, x_k$ is such that the corresponding R_i are different. This fact makes it possible to eliminate a practical difficulty that occurs in applying the classical qd-algorithm.

Just as Rutishauser's qd-algorithm, the above method can not be recommended for an accurate determination of the zeros. However, it is a good method for obtaining first approximations to the zeros, simultaneously for all zeros, using no information other than the coefficients of the interpolating polynomial. These approximations can be used as starting values for methods with faster convergence, e.g. Newton's method.

6.b. <u>Zeros of entire functions</u>.

The method just described can also be used to obtain first approximations for the zeros of entire functions, provided we have a Newton series expansion of that function.

The working scheme looks as in table 1, provided $f(x) = \sum_{i=0}^{\infty} f_{0i}\, \omega_i(x)$. From the knowledge of the first two rows, we can construct the complete scheme (as indicated by the arrows in table 1). Under certain conditions the q-columns are converging to the inverses of the (shifted) zeros s_i of f.

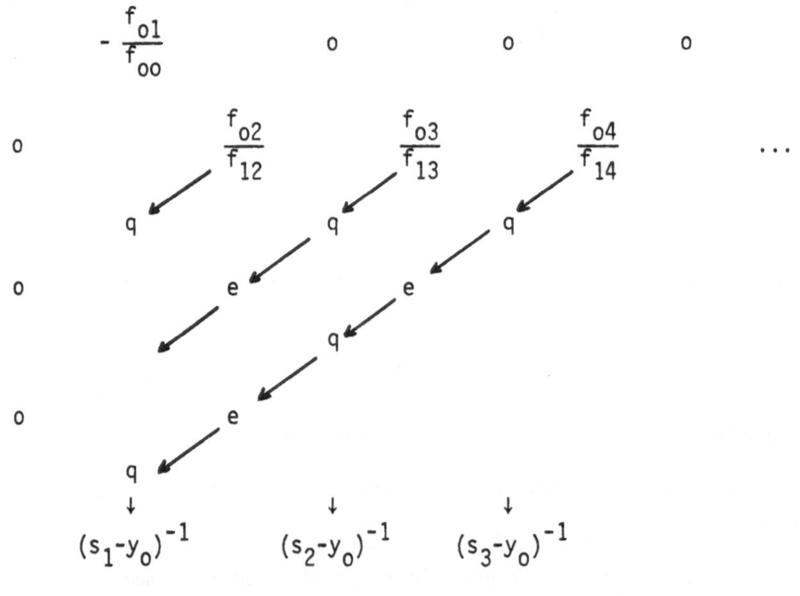

<u>Table 1</u>.

6.c. <u>Zeros and poles of meromorphic functions</u>.

Suppose we have the disposal of a Newton series expansion $\sum_{i=0}^{\infty} f_{oi}\, \omega_i(x)$ of the meromorphic function f(x). If then s_i for $i \geqslant 1$ denote the poles of $f(z)$ and z_i for $i \geqslant 1$ the zeros, then, using theorem 7 (resp. theorem 8), and provided $R_{m-1} < R_m < R_{m+1}$ (resp. $R'_{m-1} < R'_m < R'_{m+1}$) the m-th q-column (resp. m-th e-row) in the generalized qd-table converge to the inverses of the (shifted) poles (resp. zeros) of f.

The working scheme looks as in table 2.

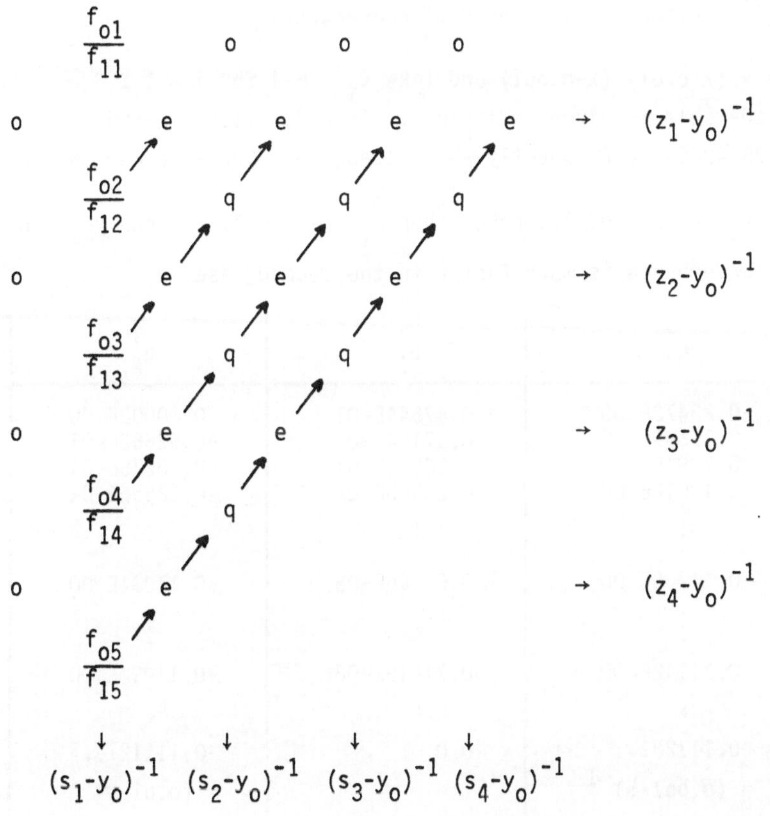

Table 2.

7. NUMERICAL EXAMPLES.

In this section we give some numerical examples to illustrate the convergence properties of the generalized qd-scheme.

Example 1.

First we want to illustrate the fact that the convergence of the q-columns heavily depends upon the choice of the interpolation points. Indeed, as can be seen in theorem 2, the choice of the limitpoints fixes the R_i and the ratios $\frac{R_{i+1}}{R_i}$ determine the speed of convergence.

Let $f(x) = (x-0.01).(x-0.001)$ and take $x_i = 8+i$ for $i = 1,2,3,4$. The corresponding qd-table is given in table 3. Then $R_1 = 11834.28586$ and $R_2 = 11875.42266$, consequently $\frac{R_2}{R_1} \simeq 1.003476$. For a second choice (table 4) we take $\{x_i\} = \{1,0.02,5,0.005\}$, then $\frac{R_2}{R_1} \simeq 1.23094$. As can be seen in table 3 and 4 convergence is much faster in the second case.

i	q_1^i	e_1^i	q_2^i
0	-0.23472E 00	0.47644E-01	0.00000E 00
1	-0.17589E 00	0.27154E-01	-0.52662E-01
2	-0.13918E 00	0.18533E-01	-0.70275E-01
4	-0.13012E 00	0.20503E-02	-0.12556E-01
	⋮	⋮	⋮
100	-0.11205E 00	0.53046E-05	-0.11031E 00
	⋮	⋮	⋮
300	-0.11142E 00	0.71419E-06	-0.11094E 00
	↓	↓	↓
	-0.11123... $= (0.001-9)^{-1}$	0.0	-0.11112... $= (0.01-9)^{-1}$

Table 3.

i	q_1^i	e_1^i	q_2^i
0	-0.10202E 01	0.19964E 00	0.00000E 00
1	-0.10202E 01	0.26935E-02	-0.99108E 00
2	-0.10065E 01	0.13476E-01	-0.99112E 00
4	-0.10078E 01	-0.25102E-03	-0.10046E 01
	⋮	⋮	⋮
46	-0.10101E 01	-0.96189E-04	-0.10011E 01
	↓	↓	↓
	-0.10101 ... $= (0.01-1)^{-1}$	0.0	-0.10010 ... $= (0.001-1)^{-1}$

Table 4.

Example 2.

With the generalized qd-scheme it is possible to find complex conjugate zeros of a polynomial. It suffices to insert one (or more) complex points x_i. Let $f(z) = z^4-8z^3+39z^2-62z+5 = (z-3-4i)(z-3+4i)(z-1-i)(z-1+i)$, and $\{x_i\} = \{3,i,-2,-1,2\}$.

In this example $R(z) = |(z-i)(z+1)(z-2)(z+2)(z-3)|$.

And, as a consequence of the insertion of a complex point, complex conjugate zeros will in general give different values of R.

A part of the generalized qd-table for this example is displayed in table 5.

The same device can be applied to find real zeros which have the same absolute value but which differ in sign.

i	q_1^i	q_2^i	q_3^i	q_4^i
1	-0.26749 - i.0.12200	-0.34124 + i.0.51093	0.01401 - i.0.01896	0.23529 + i.0.05882
2	-0.36430 - i.0.53120	-0.66252 + i.0.29453	-0.61308 - i.1.31520	0.60000 + i.1.30000
4	-0.52647 - i.0.48927	-0.27802 + i.0.48990	-0.01487 - i.0.07792	0.22568 + i.0.15708
50	-0.39997 - i.0.19996	-0.40006 + i.0.20012	0.00273 - i.0.26535	-0.00974 + i.0.26205
100	-0.40000 - i.0.20000	-0.40000 + i.0.20000	-0.00061 - i.0.25014	0.00048 + i.0.25041
	↓	↓	↓	↓
	-0.4-0.2.i	-0.4+0.2.i	-0.25.i	0.25.i
	$= (-2+i)^{-1}$	$= (-2-i)^{-1}$	$= (4.i)^{-1}$	$= (-4i)^{-1}$

Table 5.

REFERENCES

[1] BAKER, G.J. Jr. : The theory and application of the Padé approximant method. Advances in Theoretical Physics 1 (1965), 1 - 58.

[2] BARNSLEY, M.F. : The bounding properties of the multipoint Padé approximant to a series of Stieltjes on the real line. Journal of Mathematical Physics 14 (1973), 299 - 313.

[3] CLAESSENS, G. : A generalization of the qd algorithm. Journal of Computational and Applied Mathematics. To appear.

[4] CLAESSENS, G. : On the Newton-Padé approximation problem. Journal of Approximation Theory 22 (1978), 150 - 160.

[5] CLAESSENS, G. : On some aspects of the rational Hermite interpolation table and its applications. Ph.D. thesis, University of Antwerp, 1976.

[6] DAVIS P.J. : Interpolation and Approximation. Blaisdell Publ., 1963.

[7] EPSTEIN, S.T. and BARNSLEY, M.F. : A variational approach to the theory of multipoint Padé approximants. Journal of Mathematical Physics, 14 (1973), 314 - 325.

[8] HENRICI, P. : Applied and Computational Complex Analysis, Vol. 1, New York, Wiley-Interscience 1975.

[9] HENRICI, P. : The Quotient-Difference Algorithm. National Bureau of Standards - Applied Mathematics Series 49 (1958), 23 - 46.

[10] HOUSEHOLDER, A.S. : The numerical treatment of a single nonlinear equation. New York, Mc Graw-Hill, 1970.

[11] HOUSEHOLDER, A.S. : The Koenig-Hadamard theorem again. In "Studies in Numerical Analysis" (B. Scaife, ed., Academic Press, London, 1974), 235 - 240.

[12] DE MONTESSUS DE BALLORE, R. : Sur les fractions continues algébriques, Bull. Soc. Math. France, 30 (1902), 28 - 36.

[13] PERRON, O. : Die Lehre von den Kettenbrüchen. Band II. Teubner, Stuttgart, 1957.

[14] RUTISHAUSER, H. : Der Quotienten-Differenzen-Algorithmus. Zeitschrift für Angewandte Mathematik und Physik 5 (1954), 233 - 251.

[15] SAFF, E.B. : An extension of Montessus de Ballore's theorem on the convergence of interpolating rational functions. Journal of Approximation Theory 6 (1972), 1 - 5.

[16] SHAMASH, Y. : Continued fractions methods for the reduction of discrete time dynamic systems. International Journal of Control, 20 (1974), 267 - 275.

[17] SHAMASH, Y. : Linear system reduction using Padé approximation to allow retention of dominant modes. International Journal of Control, 21 (1975), 257 - 272.

[18] WALSH, J.L. : Interpolation and approximation by rational functions in the complex domain. Colloq. Publ. Vol. 20, American Math. Society, Providence, 1969.

[19] WARNER, D.D. : Hermite interpolation with rational functions. Ph.D. thesis, Univ. of California, 1974.

Authors address : University of Antwerp, UIA
Department of Mathematics,
Universiteitsplein 1, B-2610 WILRIJK
Belgium.

Unter dem Titel "Numerische Methoden der Approximationstheorie" sind folgende Bände im Birkhäuser Verlag erschienen:

Band 1 (ISNM 16)

Vortragsauszüge der Tagung über Numerische Methoden der Approximationstheorie vom 13.-19. Juni 1971 im Mathematischen Forschungsinstitut in Oberwolfach (Schwarzwald)
Hrsg. von L. Collatz und G. Meinardus
1972, 246 Seiten. Pappband
ISBN 3-7643-0633-5

Band 2 (ISNM 26)

Vortragsauszüge der Tagungen über Numerische Methoden der Approximationstheorie vom 3.-9. Juni 1973 im Mathematischen Forschungsinstitut in Oberwolfach (Schwarzwald)
Hrsg. von L. Collatz und G. Meinardus
1975, 200 Seiten. Pappband
ISBN 3-7643-0764-1

Band 3 (ISNM 30)

Vortragsauszüge der Tagungen über Numerische Methoden der Approximationstheorie vom 3.-9. Juni 1973 im Mathematischen Forschungsinstitut in Oberwolfach (Schwarzwald)
Hrsg. von L. Collatz und G. Meinardus
1976, 334 Seiten. broschiert
ISBN 3-7643-0824-9

Band 4 (ISNM 42)

Vortragsauszüge der Tagung über Numerische Methoden der Approximationstheorie vom 13.-19. November 1977 im Mathematischen Forschungsinstitut in Oberwolfach (Schwarzwald)
Hrsg. von L. Collatz und G. Meinardus
1978, 344 Seiten. broschiert
ISBN 3-7643-1025-1

Band 5 (ISNM 52)

Excerpts of the Conference on Numerical Methods of Approximation Theory, March 18-24, 1979 at the Mathematical Institute Oberwolfach, Black Forest

Hrsg. L. Collatz and G. Meinardus

1980, 338 Seiten. broschiert

ISBN 3-7643-1103-7